SMART EVALUATION AND INTEGRATED DESIGN IN REGIONAL DEVELOPMENT

In memory of my mother

Smart Evaluation and Integrated Design in Regional Development

Territorial Scenarios in Trentino, Italy

Edited by

GRAZIA BRUNETTA
Politecnico di Torino, Italy

Routledge
Taylor & Francis Group
LONDON AND NEW YORK

First published 2015 by Ashgate Publishing

2 Park Square, Milton Park, Abingdon, Oxfordshire OX14 4RN
52 Vanderbilt Avenue, New York, NY 10017

Routledge is an imprint of the Taylor & Francis Group, an informa business

First issued in paperback 2020

British Library Cataloguing in Publication Data
A catalogue record for this book is available from the British Library.

The Library of Congress has cataloged the printed edition as follows:
Brunetta, Grazia.
Smart evaluation and integrated design in regional development : territorial scenarios in Trentino, Italy / by Grazia Brunetta.
pages cm
Includes bibliographical references and index.
ISBN 978-1-4724-4583-4 (hardback)
1. Regional planning--Italy--Trentino-Alto Adige. 2. Land use--Italy--Trentino-Alto Adige. 3. Regional planning--Methodology. I. Title.
HT395.I82T737 2015
307.1'2094538--dc23

2015014512

ISBN 978-1-4724-4583-4 (hbk)
ISBN 978-0-367-66865-5 (pbk)

Contents

List of Figures		*vii*
List of Maps		*ix*
List of Tables		*xi*
Notes on Contributors		*xiii*
Acknowledgements		*xv*
List of Abbreviations		*xvii*
	Introduction: Smart Evaluation, the Aims and Stages of a Work in Progress *Grazia Brunetta*	1
1	Territorial Integrated Evaluation Methodology *Grazia Brunetta*	13
2	The Design of Territorial Integrated Evaluation *Grazia Brunetta and Emma Salizzoni*	31
3	The Experimentation of Territorial Integrated Evaluation *Ombretta Caldarice and Emma Salizzoni*	53
4	The Analytic Network Process Method for Territorial Integrated Evaluation *Patrizia Lombardi*	75
5	A Mathematical Model for Territorial Integrated Evaluation *Roberto Monaco*	97
6	Smart Evaluation: The Agenda Emerging from Experimentation of the Territorial Integrated Evaluation Methodology *Grazia Brunetta*	107
Index		*125*

List of Figures

1.1	The operational dimensions of the TIE methodology	21
4.1	The BOCR model	79
4.2	Paired comparisons matrix	81
4.3	Supermatrix	84
4.4	Main network at the micro scale	87
4.5	Sample of the risks sub-network, Val d'Adige VC (macro scale)	88
4.6	Sample of the benefits sub-network, Trento (micro scale)	88
4.7	Priorities of the sub-network benefits, Trento	89
6.1	Principles for designing territorial retail development	110

List of Maps

1.1	The 16 Valley Communities	23
3.1	The retail system in Trentino	54
3.2	Territorial retail supply: Opportunity and Threat factors in the VCs	55
3.3	Territorial retail supply: Strength and Weakness factors in the cities	56
3.4	Tourism: Opportunity and Threat factors in the VCs	57
3.5	Tourism: Strength and Weakness factors in the cities	58
3.6	Land use: Opportunity and Threat factors in the VCs	59
3.7	Land use: Strength and Weakness factors in the cities	60
3.8	Landscape values in Trentino	61
3.9	Landscape: Opportunity and Threat factors in the VCs	62
3.10	Landscape: Strength and Weakness factors in the cities	62
3.11	Local governance: Opportunity and Threat factors in the VCs	63
3.12	Local governance: Strength and Weakness factors in the cities	63
3.13	Territorial retail scenarios in the VCs	68
6.1	Physical relationships: The state of the art	111
6.2	Functional relationships: The state of the art	112
6.3	Systems of territorial scenario	114
6.4	Potential physical and functional relationships: The project	116

List of Tables

1.1	Synthesis of the TIE rationality	26
2.1	TIE themes and indicators identified on the micro and macro scales	34
2.2	The relationship between TIE indicators and the principles and strategic guidelines of the PTP	42
2.3	Description of the indicators on the micro scale	46
2.4	Description of the indicators on the macro scale	47
2.5	Calculation of the indicator A1M – 'Retail consistency'	49
3.1	An exemplifying schema of the strategies, aims and actions identified for one of the 16 VC (Val d'Adige VC)	66
4.1	Saaty's fundamental scale	82
4.2	The Random Consistency Index (RCI) for different size matrices	83
4.3	Combinatory formulas for the aggregation of the results of the structured-network model	85
4.4	Val d'Adige VC ANP overall priorities (macro scale)	91
4.5	Trento ANP overall priorities (micro scale)	92
5.1	Index of attractiveness (A_i) of the VCs	102
5.2	Number of residents, tourist presences and GLA thresholds in each VC	103
5.3	Distances in km between the GLAs of each VC	103
5.4	Comparison between the distribution of residents/distribution of residents plus tourists and steady distribution in each VC	104
6.1	The emerging TIE knowledge framework for designing the territorial scenarios	108

Notes on Contributors

Grazia Brunetta, PhD in Territorial Planning and Real Estate Market, is Associate Professor of Urban and Regional Planning at the Interuniversity Department of Regional and Urban Studies and Planning of the Politecnico di Torino. She is a member of the RESURBE program and of the editorial board of the *GeoProgress Journal.* She has been a member of the editorial board of the *Italian Journal of Regional Science* and of the elected scientific board of the Italian Section of the Regional Science Association. She is the scientific coordinator of a wide range of research dealing with institutional innovation in regional and environmental planning, evaluation of territorial policies, and urban resilience. She is the author of numerous national and international publications in those fields.

Ombretta Caldarice has a PhD in Spatial Planning and Urban Development. She is research fellow at the Interuniversity Department of Regional and Urban Studies and Planning of the Politecnico di Torino and takes part in various research projects relating to retail planning. She is a member of the RESURBE program and of the editorial board of the *International Journal of Social Science Studies*. Her research activity mainly focuses on the themes of retail policies, urban regulation, territorial evaluation, and urban resilience.

Patrizia Lombardi (PhD, MSc, BA/MA) is Full Professor of Urban Planning Evaluation and Project Appraisal and Head of the Interuniversity Department of Regional and Urban Studies and Planning of the Politecnico di Torino. She has been an established figure in the field of evaluating sustainable urban development for over 20 years, publishing widely in the subject area and coordinating, or serving as lead partner, in a number of pan-European projects related to sustainably built environments, Smart Cities, post-carbon societies and cultural heritage, including BEQUEST, INTELCITY, INTELCITIES, ISAAC, SURPRISE, PERFECTION, MILESECURE-2050, DIMMER and POCACITO. She is the Scientific Director of the UNESCO Master 'World Heritage and Cultural Projects for Development' at the ITC-ILO and a representative member of the EIT-ICT Lab of Trento Rise.

Roberto Monaco is Full Professor of Mathematics at the Interuniversity Department of Regional and Urban Studies and Planning of the Politecnico di Torino. His research activity consists in, besides other items of Applied Mathematics, the derivation of mathematical models and dynamic systems applied to regional and environmental sciences.

Emma Salizzoni has a PhD in Landscape Planning. She collaborates with the Interuniversity Department of Regional and Urban Studies and Planning of the Politecnico di Torino and with the CED PPN (European Documentation Centre on Nature Park Planning), in particular, taking part in several research projects concerning landscape planning and protected area policies. Her doctoral dissertation, 'Protected landscapes: Experimental laboratories for Euro-Mediterranean coastal landscape', was awarded the Premio Tesi di Dottorato prize by the Firenze University Press (FUP) and published in 2012 by FUP. Her main fields of research are landscape policies and nature conservation policies (protected area planning and management).

Acknowledgements

The book presents the results of two research projects financed by the Autonomous Province of Trento and developed by the Interuniversity Department of Regional and Urban Studies and Planning of Politecnico di Torino, under the scientific coordination of Grazia Brunetta: 'The application of Territorial integrated evaluation methodology for designing retail territorial scenarios in Trentino' (2011–2012), and 'The Territorial integrated evaluation criteria for planning in the Trentino Valley Communities' (2013–2014).

A special thanks to Alessandro Olivi, Vice President of the Autonomous Province of Trento and Provincial Minister of the Department of Economic Development and Work, Paolo Nicoletti, General Manager of the Autonomous Province of Trento, and Luciano Fruet from the Retail Office of the Department of Economic Development and Work. They all shared in the basic concept of TIE and supported the research activity, implementing its experimentation in the provincial planning practices.

During the research I have had the opportunity to discuss the experimentation described in the book with several colleagues. In particular, I would like to thank Alessandro Balducci, Angela Barbanente, Cristina Bianchetti, Laura Fregolent, Manuela Ricci, Gabriele Paolinelli, Franco Prizzon, Marco Trisciuoglio and Bruno Zanon for their contributions to the discussion of the first research results. I would like to thank Alex Fubini and Antonio Leone for reading and commenting on the first draft of the book.

A special thanks goes out to my colleagues on the Research Team, with whom I have shared ideas and the work processes presented here. They have supported me with passion in every phases of the research: Ombretta Caldarice, Silvia Giordano, Patrizia Lombardi, Roberto Monaco, Lorenzo Piacentino, Attilia Peano, Emma Salizzoni, and Antonio Cittadino and Francesco Fiermonte from the Laboratory of Territorial and Urban Research.

Vincent Marsicano translated the Italian texts and reviewed the English ones.

List of Abbreviations

AHP	Analytic Hierarchy Process
ANP	Analytic Network Process
APT	Autonomous Province of Trento
BID	Business Improvement District
BOCR	Benefits, Costs, Opportunities, Risks
CBA	Cost-Benefits Analysis
CI	Consistency Index
CR	Consistency Ratio
CTP	Community Territorial Plan
EIA	Environmental Impact Assessment
ELC	European Landscape Convention
GLA	Gross Leasable Area
MAUT	Multiple Attribute Utility Theory
MCA	Multi Criteria Analysis
MRS	Mid-sized Retail Shop
PA	Protected Area
PTP	Provincial Territorial Plan
RCI	Random Consistency Index
SEA	Strategic Environmental Assessment
SWOT	Strengths, Weakness, Opportunities, Threats
TCM	Town Centre Management
TIE	Territorial Integrated Evaluation
TUC	Territorial Units of Cooperation
VC	Valley Community

Introduction
Smart Evaluation, the Aims and Stages of a Work in Progress

Grazia Brunetta

Research and experimentation about evaluation in spatial planning have a long history. Evaluation and planning are two interconnected dimensions. They have been defined as 'inseparable concepts' in a well-known article by Abdul Khakee (1998: 359) on the subject which highlights their close integration with each other and retraces forms of reciprocity and connections between planning theory and evaluation methodology. Following this logic, the role of evaluation does not seem to be determined univocally but interwoven with the rationality of each approach to planning.[1] Pioneer interpretations of this by Faludi and Voogd (1985), by Guba and Lincoln (1989), by Lichfield, Kettle and Withbread (1975), by Lichfield (1996), and by Mastop and Faludi (1997) identified the evolution of various evaluation methods and highlighted the significance and role in the different planning approaches. Among these, the evolutionary view formulated by Guba and Lincoln (1989) saw the 'fourth generation of evaluation' as the achievement of evaluation understood as technical argumentation in support of a strategic process of planning that is communicative, inclusive and 'transactive' (Friedmann, 1973).

The most recent turns in scholarly discussion have brought out the fact that the action of evaluation distinguishes itself during the planning process as a technique that supports decision-making (Alexander, 2006; 2009; Lichfield et al., 1998; Miller and Patassini, 2005), becoming progressively more effective and useful the more it is integrated with the planning process. This is the feature that defines the basic characteristic of evaluation, bringing out its facet as a technical argumentation in a collaborative planning process. It clarifies that there are substantive reasons for seeking out the highest level of integration with the planning process in order to further social and institutional innovation. In this sense, evaluation is a practice that puts research into action. This type of practice is one that accompanies planners as they are putting their policies together based on a continuous learning process. An evaluation approach to issues of choosing and planning is useful as a critical and analytical tool that helps bring out positions and interests. Merely technical evaluations are not enough. It is

1 For the meaning of 'transactive planning' in the evolution of planning theory, see Friedmann (1987; 1993).

fitting to place evaluation into an inclusive institutional context where actors are actively involved in giving shape to the themes up for discussion.

From the perspective that rationality is oriented to the approach of communicative planning (Forester, 1989; 2009; Healey, 2005), it is evident that evaluation can function as ‘teaching and learning’ as it promotes a process of collective learning that supports decision-making. This is a process that is circular and dialogue-oriented, that shows how important it is to put an ‘institutional design’ (Alexander, 2005) in place through evaluation/planning actions in order to renew the rationality paradigms and forms of action in response to the profound territorial changes taking place in European regional contexts.

These brief considerations are the necessary starting points for introducing the experimentation that this book presents. *Smart Evaluation and Integrated Design in Regional Development* discusses an alternative approach to evaluation in spatial planning, an approach that was experimented with in Trentino and involves the design of territorial scenarios. Specifically, *Smart Evaluation* introduces Territorial Integrated Evaluation (TIE), an innovative evaluation methodology for designing local development scenarios, thus linking the theory and the practice of planning.

This book presents the results of the research done by a work group,[2] coordinated by this writer, from the Interuniversity Department of Regional and Urban Studies and Planning for the Department of Economic Development and Work of the Autonomous Province of Trento (APT). The research originated in the process of institutional renewal in territorial planning approaches that the Province launched in 2010.

This book presents a reflection on the experimentation we performed, focusing on the meanings, instruments, rules, and operating processes of the *meta-valuation* process set up to strengthen the integration of policies/actors/ levels of governance in the territorial planning practices in the Province. As the experimentation went on, TIE methodology took on its meaning as ‘smart evaluation’, whose characteristics can be defined in this way: (1) a process of technical and institutional learning in continuous evolution – *not static analysis*; (2) a tool to help construct policies – *its characteristics are not just analytical*; (3) a form of dialogue among institutions and between institutions and the citizenship – *not a compulsory procedure of communication*.

TIE rationality is consciously ‘weak’. This logic of limited rationality defines its distinctive elements: *argumentation, inclusion,* and *qualitative methods*.

2 The Research project ‘The application of Territorial integrated evaluation methodology for designing retail territorial scenarios in Trentino’ (2011–2012) was carried out by a team of researchers from the Interuniversity Department of Regional and Urban Studies and Planning of the Politecnico di Torino: Grazia Brunetta (scientific director), Ombretta Caldarice, Silva Giordano, Patrizia Lombardi, Roberto Monaco, Lorenzo Piacentino, Attilia Peano†, Emma Salizzoni, and Antonio Cittadino and Francesco Fiermonte from the Laboratory of Territorial and Urban Research.

Characteristically, TIE is a *process* and not a *prescriptive procedure*. It uses qualitative techniques and does not produce estimates and synthetic judgements typical of quantitative techniques. For this reason, it proposes a logic of inclusion of values that express the interests and positions at play in order to design a multidimensional map of criteria and alternatives for decisions. Such an approach to evaluation seems to prefer the highest level of integration among actors/values/criteria for choosing the levels of governance, the resources, and the actors of the process of evaluation and of planning. In this logic, the perspective of TIE rationality is significant as a preparation for a 'journey' in terms of 'strategic navigation' and not as the prefiguration of the goal to reach (Hillier, 2011). This kind of journey requires the encounter of various actors/subjects/levels of government and must, just in case, be ready to change its tack while holding firm to its criteria and clear tools for the 'navigation'.

From this perspective, the TIE experimentation outlined here shows how much 'weak' rationality in evaluation corresponds to a 'high' integration between evaluation and planning, so that evaluation becomes an element inside the construction of the planning process. As illustrated in the experimentation, the evaluation approach can play a positive role in introducing new institutional knowledge and capabilities, raising the level of cohesion, cooperation and subsidiarity in the processes of territorial planning. What is at the centre of this reflection is a space for the institutional design of evaluation as an evolutionary and multidimensional process oriented to the making of territorial policies. The smart evaluation model adopts a general idea of evaluation, demonstrating the conditions that worked in favour of its application in the Trentino context of planning. This book presents the results of the experimentation performed with the objective of putting questions up for discussion, which are still relevant, on the meaning of territorial evaluation practices in the processes of planning.

I.1 A Dynamic Institutional Context: The Autonomous Province of Trento

Historically, Trentino has been exemplary within the horizon of Italian national town planning because of the innovative framework of its regulations and territorial planning.[3] Nowadays, the Province is going through a profound renewal in its public activities. The key words in this shift in perspective are *inter-sector approaches*, *integration*, and *responsibility*. The reasons for this ongoing evolutionary process can be traced back to two basic institutional conditions:

3 The first Provincial Plan, edited by Giuseppe Samonà, was adopted in Trentino in 1964 (and approved in 1967). It was a pioneering scientific practice in the national planning system (see Giovanazzi and Franceschini, 2012; Zoller, 1983). The APT is a special case in the Italian national administrative system because a 1948 law granted it autonomous legislative power in several fields, including town planning and landscape planning and conservation.

- the renewed legislation in the system of planning initiated by the approval of the Provincial Territorial Plan (PTP) – that is, the strategic framework for territorial planning, within whose parameters the inter-municipal policies of the newly instituted Valley Communities[4] (VCs) are being defined;
- the reform of provincial retail planning legislation, conducted with the enactment of the European reform regulations and national decrees on the programming of services.[5]

In this context, the provincial government has had to face an exceptional challenge, namely, to orient the new direction of territorial retail development policies to fit the reform aims, which establish the abolition of quantitative parameters, without giving up the territorial planning and the conservation of the landscape values introduced by the PTP. Thus the provincial government has had to radically rethink its planning approach to designing the processes of retail development, keeping in mind that accepting the logic of the reform decrees (liberalization of services) does not mean giving up territorial planning. Hence the provincial administration found it necessary to refer to new knowledge paradigms and instruments of governance to renew technical competences in support of public action.

It is within this concrete perspective of planning action that the APT requested the Interuniversity Department of Regional and Urban Studies and Planning of the Politecnico di Torino research group to apply the TIE methodology, which had already been experimented in other institutional regional planning contexts.[6] The goal was that of designing a new direction of provincial policies while harmonizing the needs of territorial development with those of the conservation of landscape values. This was all to be done within the horizon of strategies

4 See the provincial laws that instituted the Valley Communities (VCs), to which that very district planning reform assigned a fundamental role in inter-municipality planning: Provincial Law of 16 June 2006, n. 3; Provincial Law of 4 March 2008, n. 1 'Urban and Regional Planning'; and Provincial Law of 27 May 2008, n. 5 'Approval of the Provincial Territorial Plan'.

5 See the 'Bolkestein Directive' 2006/123/CE and Laws of 6 December 2011, n. 201 and 24 January 2012, n. 1.

6 The methodological evaluation proposal that this writer named Territorial Integrated Evaluation (TIE) is the result of a research programme coordinated and conducted from 2004 to 2008 for Piedmont Region. The research had to do with the designing and preliminary experimentation of this methodology in some territorial retail planning contexts in the region. The process of evaluation was designed in close connection with the decision-making process of regional planning, resulting in definitions of the institutional approaches, roles and meanings. With these clearly defined, it was possible to propose and test a theoretical and operational evaluative methodology that was applied to support the planning of territorial retail development, in particular gross leasable areas (GLA). See Brunetta, 2008; 2013.

of sustainability, identity, integration, and competitiveness outlined in the PTP (see Chapter 2). From the point of view of the researchers, this request was an exceptional chance for testing what had been worked on the theoretical level about territorial evaluation. This could be done through experimentation with the practice of planning.

The application of TIE led to the designing of territorial scenarios. In these, several of the leading economic factors in the development of the territories in the Province – especially tourism and retail – are able to work together as factors for regeneration, enhancement and development, under certain circumstances, in synergy with the conservation of environmental and landscape values. The assumption underlying this experimentation is the idea that retail planning cannot go on without a territorial perspective and that this perspective includes shared qualitative criteria of urban and landscape renewal in harmony with the specific characteristics of the local territories. This exceptionally dynamic context has allowed us to experiment with some emerging concepts in spatial planning, such as multidisciplinarity, new regional development paradigms and institutional capability.

I.2 Integrated Design in Regional Development

Despite institutional innovations in the field of national retail policy, integration of the sectors of retail, territory and landscape has been weakly pursued, or even completely disregarded in most Italian regions until recently. Retail policies do not generally consider the territorial aspects tied in with settlement processes, resulting in retail settlements which are not properly localized, and which therefore have a significant impact on the landscape. As a result, it is necessary to rethink this approach to the management of retail settlement processes.

Today the global economic crisis – unprecedented in its speed and depth – has made it even more necessary for us to pursue this objective, to adopt innovative solutions for the development of territories that account for the opportunities, the institutional design and the creativity of the territories. According to this rationale, the experimentation that we present supports the reinforcement of the polycentric territorial model of Trentino, understood as a scheme of spatial organization and economic development strongly anchored in the territorial identity, which has been followed in this region ever since the origin of the planning processes launched in the 1970s.[7]

Trentino is a cross-border area characterized by a lively tourist and retail demand that needs to be harmonized with the exceptional and internationally

7 The first PTP (1967) proposed the spatial metaphor of the 'extended city' in support of a polycentric development model founded on the protection and enhancement of identity values as resources for the modernization of each territorial context (valley and mountain) in Trentino.

renowned value of its nature and landscape, such as the Dolomite Mountains, a UNESCO World Heritage Site. This context means that TIE's approach to designing local development scenarios needs to look simultaneously at different themes, such as retail, tourism, territorial infrastructures, environment and landscape. The example of Trentino has allowed us to apply and redesign the TIE methodology with the aim of setting up local development scenarios which meet both the need for economic growth (in the retail and tourist sector) and the need for landscape conservation values, which are very much being fostered at the international level.

After TIE was applied, its result was the designing of territorial scenarios where retail and tourism, in synergy with the conservation of landscape values, could become factors contributing to urban renewal and territorial planning. The proposed operating processes have to do with two specific aspects of territorial planning:

- the definition of *criteria*, *objectives* and *strategies* to bring out elements of the territorial opportunities within the framework of regional development planning;
- the definition of *proactive territorial partnerships* for integrating sector-focused policies (retail/tourism/landscape) in territorial scenario design.

The territorial scenarios were projected as multi-level and multi-sector *systems of governance*, or clusters of *guiding visions*, that could activate the projection of integrated policies for the development and enhancement of local resources and local potentialities to be articulated in the institutional design of the Province.

The scenarios that come out of TIE process are not forecasts of goals to reach but potential routes to be charted by local action, routes towards territorial enhancement. Hence the process of evaluation leads to visions, updated periodically, of the development dynamics and perspectives in each territorial context. These visions would then be supports for the territorial planning choices that the VCs must make when they are designing the Community Territorial Plans (CTPs). Following this logic, the scenarios became the *settings* for starting the encounter of various local actors in order to set up the institutional design of local territories through the working out of territorial plans. The TIE *criteria* at the basis of the construction of scenarios have become the language of each local institutional discussion (in every VC) finalized at working out its own territorial development scenario.

Within such a perspective of action, the direction of work taken up responds to the objective of supporting the process of urban renewal that was undertaken through the 'tool' of TIE methodology. This process is oriented to a decentralized approach to regional planning, where local administrations enjoy full responsibility and autonomy in making decisions about the activation of territorial development strategies addressed to the future.

I.3 From Evaluation to Institutional Learning

The territorial scenarios that came out of the application of TIE were articulated in the new regulations for the governance of the VCs and municipalities. With the approval of Resolution 1339 of 1 July 2013, the Autonomous Province of Trento translated the principles and criteria of TIE methodology into *regulatory guidelines* for orienting the planning of territorial retail development. Resolution 1339 is the destination point of the research that was conducted. Through the experimentation with TIE methodology, this research led to an up-to-date analysis of the processes of retail settlement. The research ultimately proposed dominant and prospective territorial scenarios for the VCs. The conditions and parameters identified by the research for the designing of territorial scenarios in the 16 VCs are now a multi-faceted packet of norms relating to the self-evaluation of territorial plans of the valley communities as well as to the *methodology for Territorial Integrated Evaluation.*[8]

From this point of view, the book presents the process of making new perspectives of territorial governance to be implemented through the application of TIE methodology in the VCs. In general terms, this process of innovation in the knowledge paradigms and in approaches to planning is something that becomes particularly interesting in view of the present Italian national and European legislative framework. In Italy, there is the recent national law (2014) on the governing of metropolitan cities, provinces and inter-municipal bodies.[9] In the European Union, there are new lines of orientation for the designing of the new directions of territorial cohesion policies (see, for instance, Barca and McCann, 2011; Stead, 2014). In this way, the experimentation leads us to reflect on two general issues.

The first issue has to do with binding *theory* to *action*, the aim of TIE methodology. In this, the above-mentioned scenarios are not 'closed visions' but potential routes anchored in a set of *shared criteria* which are related to an inclusive process of local governance. This is a process that is founded on proactive territorial policies by the VCs – that is, policies that are focused on to the ethical choices for the economy, environment and landscape. Our research proposal helped the APT in the essential (but not always taken-for-granted) translation of the criteria for the design of scenarios into regulatory guidelines,

8 The TIE self-evaluation must be applied 'in order to promote the sustainability of the retail settlements', in particular for the Gross Leasable Areas (GLA). See section 4.3 of the Provincial Resolution n. 1339.

9 See National Law n.56 of 7 April, 2014. Law reordering the territorial administrations defines the different types of territorial divisions ('metropolitan cities', 'provinces', and the 'union and merger of municipalities') in accordance with the following criteria: territorial strategic development, efficient provision of public services, the infrastructure and communications networks, and the implementation of the European policies.

activated with the revision of the provincial law for territorial retail planning. In addition, the proposal functioned as the engine of a radical change over the course of 2014 in the way that the VCs defined their urban-planning choices for transformation and development.

The second issue has to do with the transfer of skills in the local territories: specifically, the transfer through the task of designing territorial scenarios in the VCs. After TIE methodology proposed its model and the APT translated this into indicative regulations, the research group took up the task of *accompanying* local planners over the course of 2014 so that everyone involved was enabled to translate the multidisciplinary logic of the decision-making process into local agendas. This action of knowledge support undertaken with the VCs was significant because it used the logic and the language of TIE – criteria and scenarios – to start an encounter among the actors in designing decisions about development to include in the CTPs. Therefore this was not a classical top-down action for the launch of processes of implementing scenarios proposed by TIE. Rather, it was an autonomous action of designing territorial scenarios by local actors who have started off with TIE criteria. This is the way that an intense process of discussion was started in which, in several cases, planners came to formulate proposals for designs of urban retail renewal on the basis of cooperative agreements among different actors. This aspect is particularly interesting for Trentino, a territorial reality that had been scarcely inclined to any forms of cooperative action among the different VCs.

This book demonstrates the innovative reach of the results that have been achieved for the planning process. It highlights how planners can create conditions to begin the flow of innovations in the form of actions when there is a strong will and institutional capacity. From this perspective, the theoretical background of the experimentation consists in a neo-institutionalist approach to the development of institutional innovation (North, 1990). Here the TIE criteria and scenarios become the strategic axes for the design of regional development policies, a design that is one of the factors supporting the process of institutional innovation that was started with the VCs. The results of TIE were an essential starting point for beginning the institutional discussion and the decision-making process about the territorial retail development planning that each VC had made following the territorial plans adopted by April 2015.

I.4 The Organization of the Book

The research's theoretical results are now being put into practice by Trentino territorial authorities. The innovative importance of the results achieved should spark interest for an international audience of academics, researchers and practitioners in the fields of urban and regional planning, landscape planning, and regional economics. This book, divided into six chapters, presents a reflection on the features of TIE methodology that led to the design of the processes of governance that are now being practised in each VCs. Its purpose is that of focusing on the question of evaluation in planning and the perspective of its future, its strengths, its weaknesses, and its controversial points that are still open to debate.

Chapter 1, by Grazia Brunetta, sets down the TIE methodology, illustrating the origin, rationale and perspectives of development of this process of evaluation in the realm of the process of planning. These reflections on the rules and structure of the action of *smart evaluation* lay the foundation for taking a stance on its role, meaning and usefulness and for focusing on the innovative features that originated its experimentation in Trentino.

Chapter 2, by Grazia Brunetta and Emma Salizzoni, discusses the multidisciplinary logic of TIE and demonstrates the versatility of its methodology, which is adapted according to the requests for evaluation in the institutional context of the Province. The chapter illustrates the design of criteria and presents the technical choices made in relation to the organization of the themes and indicators of the evaluation matrix.

Chapter 3, by Ombretta Caldarice and Emma Salizzoni, goes over the first steps in the experimentation – that is, the application of the TIE matrix for the interpretation of the ongoing dynamics in the Province, the definition of the three types of territorial scenarios, the signalling of the relative points of strength and weakness, and the presentation of challenges and opportunities for the prospective evolution of the scenario for each VC.

Chapter 4, by Patrizia Lombardi, presents the design of the multi-criteria Analytic Network-Process Method model, which is applied to each territorial context in order to define the conditions of strategic feasibility of the scenario that comes out of the application of TIE criteria. The chapter presents the results obtained in relation to the weight and the links of each criterion of selection in one specific area (Val d'Adige-Trento) and discusses the ANP model's potential as an instrument for helping decision-making in designing territorial development policies.

Chapter 5, by Roberto Monaco, proposes a model for estimating the attractiveness of the APT territories. The model considers the conditions of accessibility and functional integration (retail/tourism) of each VC. The model was worked out starting out with two indexes drawn from the TIE criteria. It allows planners to estimate the possible flows of people between the available gross leasable areas. From this perspective, the implementation of the model over the whole Province would indicate potential synergy among different VCs, so that the development of retail city scenarios could involve the whole Province and not only just a single VC.

Chapter 6, the concluding chapter, by Grazia Brunetta, discusses the results of the experimentation. It presents three spheres of institutional innovation of the proposals formulated by the application of TIE. The chapter discusses crucial issues related to type of territorial governance that is being proposed. Namely, smart evaluation and governance mean smart institutions in regional planning. The chapter calls for new skills and capabilities in the planning decision-making process.

References

Alexander, E.R., 2005. Implementing Norms in Practice: The Institutional Design of Evaluation. In: D. Miller and D. Patassini (eds), *Beyond Benefit Cost Analysis*. Aldershot: Ashgate, pp. 295–310.

Alexander, E.R., 2006. Evolution and Status: Where is Planning-Evaluation Today and How Did It Get Here? In: E.R. Alexander. (ed.), *Evaluation in Planning: Evolution and Prospects*. Aldershot: Ashgate, pp. 3–16.

Alexander, E.R., 2009. Dilemmas in evaluation planning, or back to basics: What is planning for? *Planning Theory & Practice*, 10(2): 233–44.

Barca, F. and McCann, P., 2011. Outcome Indicators and Targets: Towards a Performance Oriented EU Cohesion Policy. Discussion paper, European Commission, 15 February, Brussels.

Brunetta, G., 2008. *Valutazione territoriale integrata degli insediamenti commerciali*. Firenze: Alinea.

Brunetta, G., 2013. Valutazione integrata territoriale per il governo del territorio [Territorial integrated evaluation in spatial planning]. *Scienze Regionali* [*Italian Journal of Regional Science*], 12(2): 71–9.

Faludi, A. and Voogd, H. (eds), 1985. *Evaluation of Complex Policy Problems*. Delft: Delftsche Uitgers Maatschappij.

Forester, J., 1989. *Planning in the Face of Power*. Berkeley, CA: University of California Press.

Forester, J., 2009. *Dealing with Difference: Dramas in Mediating Public Disputes*. Oxford: Oxford University Press.

Friedmann, J., 1973. *Retracking America: A Theory of Transactive Planning*. New York: Anchor Press.

Friedmann, J., 1987. *Planning in the Public Domain: From Knowledge to Action*. Princeton, NJ: Princeton University Press.

Friedmann, J., 1993. Toward a non-Euclidian mode of planning. *Journal of the American Planning Association*, 59(4): 482–5.

Giovanazzi, S. and Franceschini, A. (eds), 2012. *Bruno Kessler e il primo Pup 1961–1964*. *Sentieri Urbani*, 8, Trento: INU Trentino.

Guba, E.G. and Lincoln, Y.S., 1989. *Fourth Generation Evaluation*. Thousand Oaks, CA: SAGE Publications.

Healey, P., 2005. *Collaborative Planning: Shaping Places in Fragmented Societies*, 2nd edn. London: Palgrave Macmillan.

Hillier, J., 2011. Strategic planning as strategic navigation. *CRIOS*, 1: 25–42.

Khakee, A., 1998. Evaluation and planning: Inseparable concepts. *Town Planning Review*, 69(4): 359–74.

Lichfield, N., 1996. *Community Impact Evaluation.* London: University College Press.

Lichfield, N., Kettle, P. and Withbread, M., 1975. *Evaluation in the Planning Process.* Oxford: Pergamon Press.

Lichfield, N., Barbanente, A., Borri, D., Khakee, A. and Prat, A. (eds), 1998. *Evaluation in Planning.* Dordrecht: Kluwer Academic Publishers.

Mastop, H. and Faludi, A., 1997. Evaluation of strategic plans: the performance principles. *Environment and Planning B. Planning and Design*, 24: 815–22.

Miller, D. and Patassini, D. (eds), 2005. *Beyond Benefit Cost Analysis.* Aldershot: Ashgate.

North, C.D., 1990. *Institutions, Institutional Change and Economic Performance.* Cambridge: Cambridge University Press.

Stead, D., 2014. The rise of territorial governance in European policy. *European Planning Studies*, 22(7): 1368–83.

Zoller, N., 1983. *Il Comprensorio nella Provincia di Trento. La lunga marcia verso una nuova istituzione.* Trento: Edizioni Cooperativa Culturale Rosselli.

Chapter 1
Territorial Integrated Evaluation Methodology

Grazia Brunetta

1.1 Aims of the Evaluation Process

Territorial Integrated Evaluation (TIE), the methodology being proposed and tested here, primarily aims to go beyond a merely one-dimensional approach to the regional planning of retail territorial development. In order to do this, it furnishes planners with information, thus laying more a solid foundation for the planning and design of territorial development scenarios in which the retail policy[1] could play on a convincing role in the enhancement of urban areas. The Department of Economic Development and Work of the Autonomous Province of Trento (APT), handling retail and tourism, can use this institutional and technical tool as part of their decision-making process in regional policy. TIE methodology is not meant to take the place of compulsory assessment procedures, such as Environmental Impact Assessment (EIA) and Strategic Environmental Assessment (SEA). Rather, it is meant to be a *voluntary act of institutional evaluation* to help in decision-making for planning and designing territorial transformation and development.

This entire initiative comes out of the conviction that Italy needs to change its approaches to evaluation in spatial planning. In other words, evaluation is something that is significant not merely as a technical-scientific procedure subject only to rigid regulations (such as the SEA). Rather, evaluation requires something more than, obviously, a solid supportive foundation in regulations. It requires that there be a thorough administrative and technical capacity to reorganize the practices of the stakeholders who take part in the processes of territorial planning.

Evaluation is even more worthwhile when it is interpreted as one of the many potential strategies to support the procedures of governance, so that decisions can be made that are more readily accepted and the spatial planning process can be strengthened (Alexander, 2006a; Lichfield et al., 1998; Miller and Patassini,

1 See the new territorial policy in Trentino, launched with the approval of the Provincial Territorial Plan (PTP, 2008) and Provincial Law no. 17/2010, which implements the latest European Directive (2006/123/CE) on the services and pertinent National Reform Decrees (no. 59-26/3/2010; no. 201-6/12/2011, and no.1-24/1/2012).

2005).[2] Thus strategic evaluation takes on the role of a technical process of learning accompanying the governance of decisions (Alexander, 2006a; Lichfield, 1996). It is not only a kind of dialogue conducted in order to reach decisions but also a process through which institutions can learn and accumulate more and more 'usable knowledge' (Lidblom, 1959; 1979).

Theoretically, TIE is a scientific methodology that builds innovative technical knowledge, supports local planning processes and innovates territorial governance, aiming ultimately to link regional development policies to different levels of planning (Archibugi, 2007; Brunetta, 2006; Khakee et al., 2008). Accordingly, TIE assumes that there is a general idea of evaluation in planning, and proposes this idea, one that articulates what the strategic dimension of evaluation means when it is applied to regional planning processes in order to strengthen the institutional design of the evaluative processes in the territorial planning, thus fostering innovation (Alexander, 2005). TIE is a process of *meta-evaluation* intended to act as a new capacity-building institutional framework for various territorial policies and plans. TIE's main strategic action is that it emphasizes technical communication as a learning experience for institutional authorities, seeking out the highest possible integration of evaluation with the decision-making and territorial planning process. The rationale for this kind of evaluation is the need to construct, make specific, and reach an agreement beforehand on the relative importance of the various criteria used in the decision-making process.

TIE aims to work out an overall framework of shared principles and criteria for planning territorial development scenarios in Trentino. This would be a framework to refer to when proposing cooperative agreements and stabilizing existing ones, and when defining satisfying and profitable aims for all the

2 The issue of the integration of evaluation with planning emerged in the pioneering studies in the 1970s as the essential factor for guaranteeing consistency and effectiveness to the spatial planning (Faludi and Voogd; 1985; Lichfield et al., 1975). The discussions and experimentation about evaluation and planning led, for example to the first proposal for an EU Directive (1998) on 'the evaluation of the effect of programs and plans on the environment'. From this awareness, it emerged clearly that evaluation was a strategic action and that the more it is interwoven with the process of planning the more effective it is. With the premise that the aim of evaluation is to strengthen the planning and decision-making process, territorial evaluation is really 'local'. Hence it depends on the 'territory' it is dealing with. It cannot be an unequivocal and pre-set method. Substantially, it must depend on the rational perspective of the planning context. The processes of evaluation changed over the twentieth century as a consequence of the evolution of the planning theory and of the evolution of various approaches in the European planning systems. In relation to this, Khakee (1998a) distinguishes eight 'planning theory positions' corresponding to eight approaches of evaluation. What comes out of this is an articulated and pluralistic framework that highlights the close and effective bond between 'evaluation' and 'planning', which is the source of the rationality, the meaning and the aim of the evaluation.

stakeholders involved in decision-making. On the practical level, TIE integrates evaluation into the system of territorial planning as an institutional learning process. TIE shares the open and performance characteristics of a limited rationality approach and therefore requires new methods that can reveal the strategies that are competing to use certain resources and identify which actions are most effective for making decisions. This approach helps orient subsequent decision-making phases (*in itinere* and *ex-post*) and guarantee stability at the first level of cooperation that had been reached. Over time, it can increase consensus by making its positive results visible, leading to other potential forms of agreement. Such evaluation methodology also lends to heighten the level of participation and responsibility among the various parties involved in making decisions, which become the product of a process of social interaction (March and Olsen, 1989). Each party thus participates in the process of value definition and becomes part of an organizing 'construction', which is a fundamental resource for the planning of territorial development strategies.

TIE necessarily overturns the classic paradigm of evaluation, where 'the evaluation depends on the values', assuming on the contrary that 'the values depend on the evaluation' (Archibugi, 2006: 77). That values depend on evaluation is a principle, one that derives from the meaning of evaluation as a process aimed at attributing social values, which have no fixed or intrinsic meanings.[3] Values do not depend on the objects in themselves, but take on variable properties over time that depend on socio-economic, territorial, institutional and regulatory factors – that is, the conditions, opportunities, circumstances and expectations within the territorial system (Alexander, 2009; Brunetta, 2013; Brunetta and Voghera, 2008; Moroni, 2006).[4] In short, 'the consensual valuing construction that is the outcome of the evaluation subject to continuous reconstruction (change) including refinement, revision, and, if necessary, replacement' (Guba and Lincoln, 1989: 263).

Therefore *values, evaluation* and *decisions* are limited and partial actions that come out of the specific circumstances of the issues and objects being evaluated. These are issues and objects that, in turn, come out of specific social and institutional contexts made up of precise resources, opportunities, questions and expectations that change over time. It follows that TIE helps the values that are held to be 'stable' come forth, values that can be redefined and designed in order to support regional policy and territorial planning choices. The core of

3 The values can be appreciated only in the decision-making process; their validity depends on the process itself. Here evaluation plays the role of a technical learning process inside the decision-making process whose goal is to make explicit the values and the criteria used for making public choices. In this sense, 'the values depend on the evaluation' and '… no portion of the evaluation process can be considered to be value-free'. (Guba and Lincoln, 1989: 263).

4 For a critical review of the ethical positions of the evaluation approaches in planning, see Alexander (2006b), Khakee (1998b) and Moroni (2006).

the TIE approach to evaluation is the construction of a shared knowledge and technical framework through the re-composition and progressive integration of resources, selection criteria and stakeholders.

Evaluation is designed in planning while it is being done. Afterwards, it is updated incrementally, as answers are being sought out to the problems that are arising in the course of the action. This strategic approach to evaluation is focused on the design of a shared technical and knowledge framework.[5] This approach introduces meta-criteria that progressively recompose the positive effects produced over the course of the process of the making the decisions, meta-criteria that promote potential new forms of agreements (March and Olsen, 1995). The application of TIE in Trentino shows that meta-criteria – such as sustainable development, bio-based economy, and green and integrated transport – play roles that adopt themselves according to their positions during the act of evaluation in the planning process in order to help lay out the path of planning and motivate the direction it takes in the various territories of the Province (see Chapter 4).

TIE articulates the strategic logic of evaluation as a process with three main features:

- It is a *gradual learning* process that is continuously evolving. As such, an integral part of TIE is the design of improved monitoring systems that follow the implementation of territorial development;
- It is oriented towards the *construction of policies* (it is not just analytical). It is a means for innovating design, verifying effectiveness during the implementation phase, correcting errors and improving performances;
- It is a *form of dialogue among institutions and between the institutions and citizens*, the aim of which is to improve the level of cooperation and subsidiarity.

We should think of evaluation as a knowledge framework inside which we can go on to 're-construct' the interrelations among resources, stakeholders and strategies progressively. The strategic logic of evaluation and spatial planning is something that goes beyond the technocratic dimension of the act of evaluation. Instead, this logic includes an approach directed towards proactive institutional learning that is initiated through a tailor-made approach that adapts itself to the planning approach in each territorial context.[6] TIE plays a role in guiding and designing planning process based on a continuous institutional learning

5 See the Community Impact Evaluation by Lichfield (1996), which proposes the evaluation issues both from a technical and a communicative perspective to manage complex values in planning choices as use values, non-use values, existence values, intrinsic values and also market and cost values.

6 Evaluation is the introduction of a technocratic element in a political decision-making process. Thus we should interpret the strategic dimension of the action of evaluation

process. In this sense, the process of defining principles and values becomes an indispensable step in the decision-making process. In fact, it could facilitate the 'strategic navigation' of future development trajectories in spatial planning[7] (Hillier, 2011).

1.1.1 The Meaning of Integration

The process of TIE in designing territorial planning scenarios for retail settlement policies is accompanied by the procedures of legally mandated evaluation. A primary example of these is the EIA, which is made during the final phase of the decision-making process and oriented mainly towards checking whether the project conforms to the planning aims. TIE goes beyond this prevailing logic of evaluation. It is not concerned with estimating the specific impact of any specific settlements project. As a result, this integrated territorial evaluation approach does not concern an estimate of the specific impacts determined by a precise settlement project (as in the case of the EIA).

TIE has a different emphasis. It focuses on combinations of the cumulative *territorial effects* of the potential retail development processes of Gross Leasable Areas (GLAs). In dealing with these effects, TIE is conceptually and operationally different from Cost-Benefits Analysis (CBA) and EIA, the two assessment methods for retail settlement projects. CBA uses a variety of countable evaluation techniques in order to propose a quantification of the various impacts. EIA uses market values rather than estimates of costs to relate the kinds of environmental impact that it identifies, proposing qualitative estimates of these impacts. EIA aims to check whether a project conforms to planning restrictions in order to estimate the extent of any adjustments to be

---

as a framework within which planners can proceed to a progressive 'reconstruction' of the interactions among the resources involved in the play of the decisions.

7 In particular, Hillier (2011) proposed an effective metaphor – navigation – that pertains to the dimension of uncertainty in spatial planning. Spatial planning as navigation may be a metaphor more fit to apply to such complex systems. Nevertheless, this approach must always be open to changes according to the circumstances. So, it must be able to change course when facing new problems and new chances to reach the aims that have been plotted out. Spatial planning as navigation requires a decision-making process that can be discussed by its participants, checked and updated. This requires that several elements be defined in advance – the direction of the voyage, the destinations to reach, and the tools from which to depart. TIE assumes this theoretical perspective as a voluntary information-gathering technical tool for supporting and orienting planning. The aim of TIE is to define the principles and conditions at the basis of the aims that are meant to be reached, initiating (with the navigation) the design of planning actions in local territories. This 'journey' requires that the planners, the citizens affected by the planning, and very different various levels of government get together. Thus this process must have the same characteristics as navigation – clear and shared opinions in the mind – and must be ready to adapt itself to change routes if unexpected circumstances arise.

made to protect environmental resources. Chiefly, both CBA and EIA estimate the impact – CBA, economic, and EIA, environmental – of establishing a new activity, using variables and values that are assumed to be definitive. Both CBA and EIA are based on the assumption that it is already clear to everyone involved what the resources at stake are – environmental, territorial and landscape-related – and that these resources will end up being used only at the end of the decision-making process, after the framework of the project's variables (subject to assessment) has been defined.

For CBA and EIA, assessment is essentially a compliance check based on the assumption that the evaluation criteria and their relative importance have already been defined within the context of the general aims of the territorial planning choices. This sequence ideally traces a model of abstract decisional rationality. It is, however, broadly recognized that the process described above is not implemented in the planning practice according to a linear framework linking knowledge, decision and action. Rather, what usually happens is that public actors interact and construct agreements that result in setting up the relative ranking of the criteria for decision-making. These rankings are not a priori data. In the step from the regional policy level to that of local planning, it will become necessary to translate the framework of conditions and criteria to support the design of shared development choices in the various areas of Trentino.

When planners are trying to make decisions about establishing GLAs, it is impossible for them to follow linear/sequential hierarchy models that have been established in advance and pay no attention to the specific nature of each context. Rather, they present, compare and contrast various points of views and expectations.[8] Significantly, TIE takes an approach that is not a neutral one, as the other approaches allege to be. TIE, instead, considers the ethical perspective in designing the process. TIE emphasizes that there is no unilaterally determined solution. In other words, there is no a priori solution that can be conveyed through allegedly neutral evaluation approaches, such as estimating the sum total of the individual preferences. Rather, a solution can also be the result of sharing principles and criteria that are not a priori, but emerge from the evaluation process itself (Lichfield, 1988; 1996). In this way, TIE can become an instrument that helps construct the values needed for decisions. Planning, of course, involves sector-focused decisions, those that are based on the priorities and points of views of a particular sector of society or the economy. From the perspective of TIE, these sector-focused decisions may be valid, but need to be compared, assessed and made to be consistent with a multidimensional point of view (Brunetta and Peano, 2003; Khakee, 1998b; Lichfield et al., 1998; Miller and Patassini, 2005a; Palumbo, 2001; Patassini, 2006; Voogd, 1998).

In this view, Alexander (2005: 296–7) also suggests the following:

8 On the need to consider experimentation in strategies and techniques for identifying, measuring and incorporating non-market values in planning evaluation methods, see Miller and Patassini (2005b).

> the main problems with plan and project evaluation today cannot be solved with consciousness-raising methods: they demand deliberative and effective institutional design. Current planning paradigms give planners, evaluators and analysts a scientifically objective position or a socially benign role in a neutral evaluation process. The norms are demonstrably incompatible with our actual life-world [...] Reality is different: evaluation methods and frameworks are weapons in a planning process that is a partisan debate between conflicting interests, and evaluation is usually a contested, non a consensual process.

Within the logic of planning GLA settlements, the decision-making process cannot but start off from an *ex-ante* evaluation of the current territorial retail dynamics. TIE assumes that it can represent the proposals being put forward adequately, which refer to the status of retail opportunities being made in specific territorial areas that are available. Therefore TIE can represent the issues to consider in the making of new retail development scenarios in Trentino. TIE suggests the use of qualitative evaluation techniques that will flank the consolidated assessment procedures with a process of scenario construction.

TIE's approach aims to foster a maximum level of connection with stakeholders/values/selection criteria/evaluation criteria. TIE's strategic dimension is characterized as an institutional *learning process*, and not just a *technical procedure*. As we will see, it uses qualitative methods, such as SWOT (Strengths, Weakness, Opportunities, Threats) and multi-criteria analysis methods for the making of scenarios (see Chapter 4). It does not produce the static synthesis of estimates and judgements typical of quantitative methods like CBA. TIE proposes a rationality framework for the inclusion of social values (which express the interests and the positions at stake) in order to design a multidimensional map of criteria as well as to prepare alternative decisions about territorial development scenarios, with a particular focus on new retail settlements.

The aim of TIE is to consolidate and progress beyond the first level of consensus, assuming that the decisions about new retail development should be the outcome of a certain level of consensus. Decisions should be made in view of the plurality of questions and interests expressed by the stakeholders within the Province. These are not only the interests expressed by the stakeholders promoting the development of the retail sector, but also the interests expressed by other stakeholders as well as by the potential users and citizens.

In this sense, the TIE methodology, which originated in the application of retail regional planning, is a 'tool' for facing the multitude of potential phenomena generated by establishing GLAs. The establishment of GLAs has several effects – features related to the territorial dimension, the economy, the environment and the landscape. These features set off processes that are cumulative and synergetic, seemingly intensifying the space and time involved.

As a 'tool' for facing these phenomena, TIE is particularly important for GLA settlements because these settlements can generate transformations that affect vast parts of the territorial context and set off cumulative and synergetic processes among different types of potential territorial effects. Therefore the TIE approach tries to identify potential causal relationships among effects that are 'direct' and 'indirect'. The direct effects are those that essentially relate to an evaluation of the retail activity – that is, consistency and typology in the existing retail supply, and employment in the retail sector. The indirect effects are those that relate to intangible effects – that is, those that impact territorial planning and local governance.

The localization of the retail settlements calls for the following interwoven dimensions of analysis and evaluation:

- territorial – in relation to the availability and the state of soil sealing;
- landscape – regarding the state of environmental resources and ecosystems as well as the cultural assets of the local landscape;
- economic – concerning the effect on the economic functions, the impact on local territorial systems and the potential relationships between existing and planned economic activities;
- social – relating to models of consumption, emerging lifestyles and the quality of the retail system.

The TIE methodology is based on this organization of these concepts of analysis and evaluation. It concentrates on an in-depth analysis of the first three inherent evaluation dimensions:

- *territorial* (land use and infrastructure);
- *landscape* (the state and the dynamics that concern environmental and cultural resources);
- the *local economy* (characteristics of the supply and demand in retail and tourism).

The aim of TIE is to go beyond a sector-focused approach to retail planning, the consolidated approach in Italy. Instead, TIE proposes the working out of spatial planning scenarios. In these, retail loses its strictly sector-focused value and becomes a potential factor, where circumstances allow, in promoting actions of urban renewal and landscape enhancement.

As we will see later, TIE does not limit itself to checking whether the aims of a sector have reached certain levels, for example, whether a system of territorial retail supply is being implemented or has been completed. TIE analyses the dynamics in progress in the local territorial system, and evaluates the potential effects of the new retail development on local resources. Because TIE is an evaluation methodology that is integrated, it deals with the entire combination of potential territorial effects on five interconnected evaluation themes: Territorial retail supply, Tourism, Land use, Landscape, Local governance. These effects

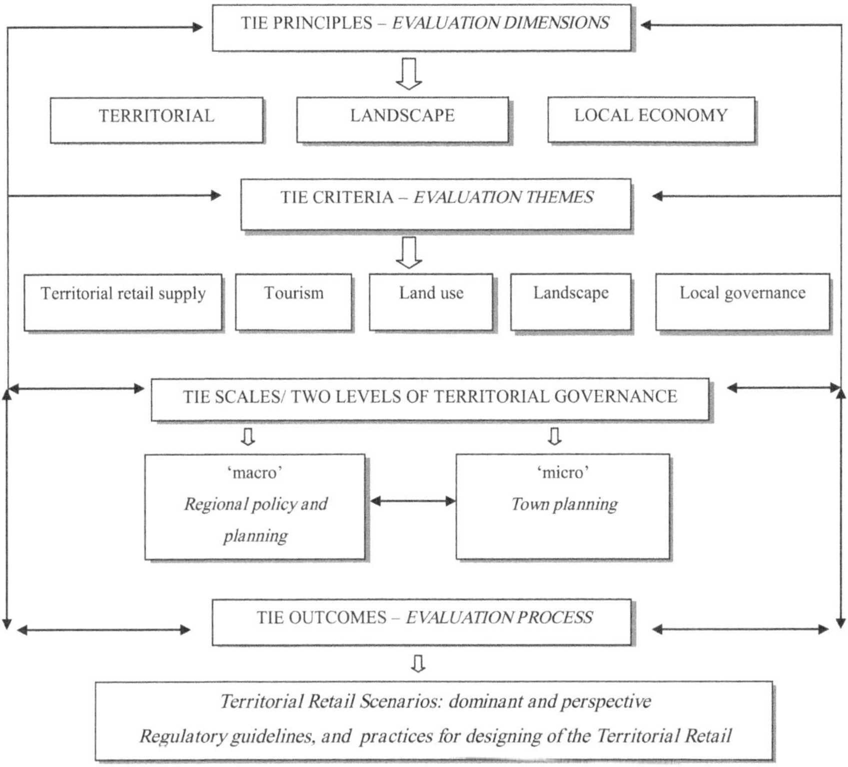

Figure 1.1 The operational dimensions of the TIE methodology

Source: Author's elaboration.

are the ones that are usually dealt with on a sector-by-sector basis by standard evaluation procedures (Figure 1.1).

In this sense, TIE becomes a 'tool' inside the decision-making process, giving planners a chance to promote and strengthen the institutional paths of spatial governance and evaluation. Doing this, TIE has an approach that is strategic because it depends on the performance of those involved. As performance, this approach follows a theoretical rationality paradigm that is consciously 'weak' – in other words, logic that does not claim to have reached any strong, solid truth. This 'weakness' seemingly helps TIE tackle two needs successfully: first, the need to strengthen the effectiveness of evaluation in spatial planning; and, second, the need to strengthen environmental assessment procedures so that TIE can interact more closely with the other assessment procedures, improving their quality and usefulness.[9]

1.2 The 'Rules' of the Evaluation Process

TIE methodology is structured as a consistent, logical and rational process with specific procedures, tools and techniques of analysis and evaluation for the design of territorial development scenarios. TIE is an institutional process that is 'self-organized' and holistic. Thus TIE is not a compulsory procedure, but a technical process of institutional learning. The fact that it is a learning experience does not mean that there are no protocols for the evaluation action. Clearly, TIE has an operational dimension that deals with the resources actually available in the institutional system of evaluation and planning. From this perspective, first and foremost, TIE requires institutions able to reorganize the practices of public actors so that they can progressively integrate skills at the different levels of territorial governance, skills that are weakly connected and usually fragmentary, redundant and dispersive.

The design of the 'rules' of TIE takes into account the local features of the territorial system (subject to evaluation), proposing the inclusion of different local resources in the grid of values that are to support the emerging territorial scenario proposals. This is a substantial upending of the role of evaluation in territorial planning; it proposes a switch from a centralized approach to a more 'inclusive' one and is aimed more at launching a joint and strategic decision-making process for designing future *development scenarios* within Trentino's Valley Communities (VCs).

TIE answers the call to propose new knowledge tools and techniques for the Trento Provincial Department of Economic Development and Work. The 'tools' consist in the process of choosing and working out concise indicators. The techniques consist in the geographic representation of phenomena, evaluative matrices, maps of strengths and weaknesses, and maps of dominant scenarios and prospective ones. These tools and techniques were worked out from freshly collected data in the data banks of the Autonomous Province of Trento, which is being updated periodically. These instruments and techniques are meant to make it possible to evaluate the territorial retail dynamics *ex-ante* and to monitor them successively – that is, the dynamics that come out of the decision-making processes that the provincial administration would initiate with the VCs. This

9 For example, it can be useful to consider the possible interactions with the compulsory assessment procedures, such as EIA and SEA. In this sense, TIE and compulsory assessment procedures can coexist and affect one another, strengthening the validity of the evaluation process, and thus helping to strengthen the decision-making process of territorial planning.

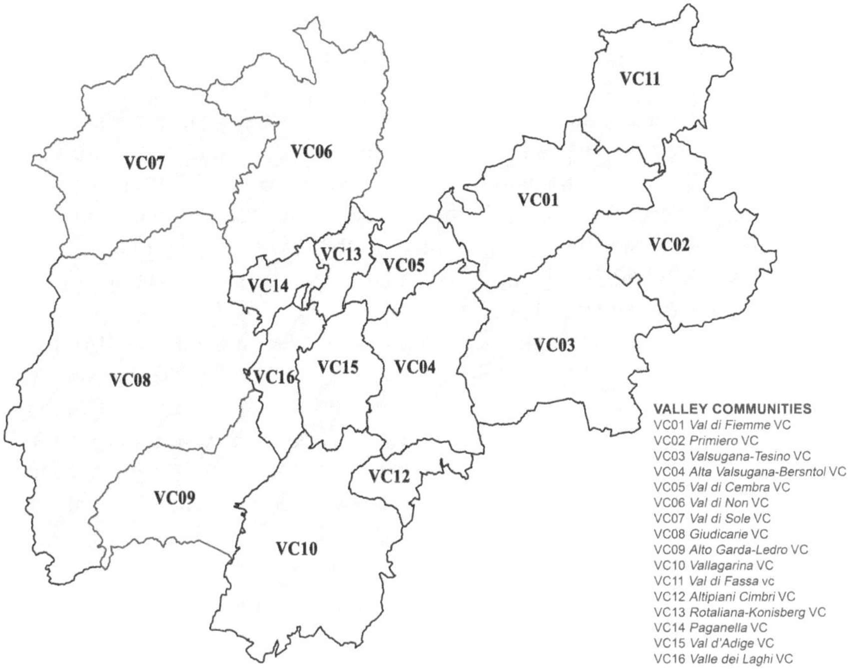

Map 1.1 The 16 Valley Communities

Source: Author's elaboration.

is an emphatic example of the versatility of TIE, which can adapt itself to the specifics of the territorial context that is the object of evaluation.

These changes mark an ongoing process of renewal in the way the process of territorial retail planning is being organized in Trento Province. Around these changes, TIE has outlined principles, conditions and criteria for orienting the new territorial development perspective. The perspective on planning that the Province intends to initiate is one that has to be designed entirely, one that is harder to put into effect, but one that is certainly more promising in terms of the quality of the transformations that are meant to be put in effect as the result of the decision-making process.

Following this rationality, TIE has set up these 'rules', which apply to the Trento Province for the planning of new GLAs, rules that introduce innovative transitions:

- from mono-sector evaluation to multi-sector evaluation;
- from a vision of regional policy centred on single decisions/projects to one that focuses on the territorial system;
- from a list of authorizations to a territorial agenda of scenarios;

- from objectives of compensation and mitigation centred on single projects to integrated scenarios of territorial enhancement.

The result of the application of TIE is the definition of territorial retail scenarios, articulated into a set of non-prescriptive spatial strategies and lines of direction for the designing of the planning process in Trentino's VCs. These scenarios were to be discussed and shared in the local decision-making process that each VCs had to implement in 2015 to respond to the new territorial planning tasks that they were assigned to by the Province in regard to spatial planning. In fact, the criteria for the territorial scenarios that came out of TIE were spelled out in the new provincial regulations. The provincial administration's Resolution 1339 of 1 July 2013 transformed the principles and criteria of TIE methodology into regulatory guidelines for orienting territorial planning and urban retail design in the VCs and municipalities. The conditions and parameters identified by the research for the designing of scenarios in the 16 VCs have become a multifaceted set of norms concerning the retail development in the Community Territorial Plans (CTPs).

1.3 The Structure of the Evaluation

To organize the structure of the evaluation, TIE adopts and adapts some of the elements of the multi-criteria and SWOT analysis methods, making several adjustments and changes (see chapters 3 and 4). Specifically, the structure of TIE is based on these three principles:

- Reconstructing the knowledge framework of the different categories of potential effects generated by retail development in Trentino. The nature of retail settlements generates various kinds of effects that spread over the territorial system. This development affects not only the structure of the retail territorial supply but also other components in relation to the inhabitants in each VC. For example, the effects on the chances of local actors to take initiatives in other leading economic sectors of the territorial context, especially tourism and agriculture, the primary forces in the economy of Trentino. TIE defines a territorial vision of this set of potential dynamics induced by retail functions and proposes the development of spatial retail policies that give local actors a chance to promote urban renewal and the enhancement of territorial landscape;
- Launching systematic evaluation of planning on the territorial level as a way of supporting decision-making in relation to the development of public sectors in Trentino, with a special focus on GLAs and their integration with other economic functions, such as tourism. In this, TIE plays a supporting role in decisions about territorial development planning. TIE aims to highlight the ongoing settlement dynamics and

thereby identify territorial development scenarios that can promote actions leading to renewal and enhancement policies. TIE defines a qualitative, multi-scale and multi-sector evaluation approach to the monitoring of the retail dynamics in the Province;

- Adopting sustainability as a criterion for designing territorial scenarios that conform to conditions and strategies defined by the Provincial Territorial Plan (PTP), keeping in mind that these conditions and strategies need to be compared and contrasted with the structure of integrated evaluation that has been designed. TIE defines strengths, weaknesses, opportunities and threats in order to identify dominant and prospective scenarios for each VC, scenarios that aim to integrate enhancement and renewal in the emerging territorial systems, not only including the enhancement of the quality of the retail sector.

These three principles aim to design regional policies that can provide a combination of guidelines, conditions and criteria that promote steps to take in local planning to strengthen, diversify and enhance the territorial well-being, all these on the local level. TIE helps planners identify 'potential' actions for the improvement of each VC. Such identification consists in the laying down of the first non-prescriptive criteria addressed to institutional and retail stakeholders. These types of criteria *work to shape a set of values* that can be used for focusing on competition, on especially critical issues, on the eventual overlapping of competencies, and on potential interactions. This focus may facilitate one or more of the scenarios that come out of TIE (Table 1.1).

1.3.1 The Two Governance and Planning Levels of the TIE Process

As an ongoing action that is put into operation, the TIE model begins to operate by fixing the identity of the institutional context that will be handled and the range of the territorial level of evaluation. Thus TIE, in operation, has become a method that works concretely as a tool for helping in local decision-making processes. In accordance with the theoretical evaluation logic described above, the operational dimension of TIE is characterized by its communicative, inclusive and 'local-based' position strongly linked to the territorial context (Guba and Lincoln, 1989).[10]

10 In this sense '[...] evaluation is a local process. Its outcomes depend on local contexts, local stakeholders, and local values and cannot be generalized to other settings' (Guba and Lincoln, 1989: 253).

Table 1.1 Synthesis of the TIE rationality

Scales	Policy/Planning
Levels of governance	*Regional/Inter-municipalities (Valley Communities – VCs)*
Actors	*Autonomous Province of Trento/VCs/Municipalities/ Stakeholders*
Evaluation approach	*Performance-based*
Operational dimension	*Gradual institutional learning/construction of policies/form of dialogue among institutions/communicative local-based process between conflicting positions*
General criteria	*External coherence (with the principles of the Provincial Territorial Plan)*
Specific criteria	*Local economy/Land use/Landscape and cultural heritage*
Conditions	*Territorial accessibility/urban sustainability of the retail development policies*
Evaluation methods	*SWOT and ANP Analysis*
Output	*Norms and regulatory guidelines/non-prescriptive criteria/ practices for designing of the Territorial Retail Planning*
Spatial outcome	*Territorial Retail Scenarios: dominant/prospective Identification of potential actions for the territorial improvement of each VC*
Time scale	*Mid-term and long-term*
Type of interaction	*Institutional inclusiveness at all levels Participation process – Public discussions and workshops*
Results	*Institutional technical knowledge framework to support the designing of shared development choices in the VC's planning process*

Source: Author's elaboration.

TIE defines the 'scales',[11] the themes, and the parameters of evaluation process on the *two levels* of territorial governance it operates on – *provincial/* territorial policy and planning and *municipal/*town planning.

TIE is applied first on the institutional level of provincial territorial policy and planning. This is the level that provides the institutional framework for initiating any comparisons of TIE with current local development policies. These comparisons would reveal potential qualitative improvements and the needs for territorial enhancement. The evaluation model proposes an up-to-date knowledge framework of Trentino's dynamics so as to sustain the network's potentially virtuous effects on the territorial system as a whole via local revitalization and enhancement strategies and objectives.

From this perspective, the structure of the TIE process in the Province involves two interrelated 'scales' of evaluation, which correspond to two levels

11 In the meaning of the level of territorial planning involved in the evaluation process.

of territorial governance and planning, each with its own precise mandate, purposes and resources:

- macro-territorial evaluation, aimed at designing a spatial planning scenario for each VC of the Province;
- micro-territorial evaluation, aimed at designing town-planning guidelines with particular focus on the five municipalities with more than 10,000 inhabitants.[12]

On the 'macro' scale, TIE enables planners to define a dominant and a prospective position for all 16 VCs on the basis of meta-criteria and specific criteria, in order to orient provincial development scenarios for the present and the future in each territorial context (inter-municipal and urban).

In particular, evaluation at the 'macro' scale performs the following actions:

- choosing the evaluation criteria and setting the indicators of TIE evaluation matrix;
- identifying how qualified the VCs are, in view of the indicators of the five TIE themes and evaluating the opportunities and risks of each VCs;
- constructing the ranking among the TIE evaluation criteria;
- positioning each VC in relation to an index of attractiveness;
- positioning each VC in respect to the three territorial scenarios ('Retail – Designing the territorial retail system'/'Marketing – Designing the integration of retail, tourism and landscape'/'Landscape – Enhancing the landscape identity of the territorial system');
- identifying systems of relationships among VCs in support of a proposal of geometrically variable systems of territorial scenarios.

Micro 'scale' evaluation is an evaluation to be applied on a local level. Its aim is to bring the potential of initiating emerging territorial scenarios for the 16 VCs together with the ongoing dynamics in the main urban areas of the Trentino. On a 'micro' scale, TIE answers the question about the definition of the positioning of urban areas and the relative roles that they could play in the territorial scenarios that were identified. Here TIE takes on a meaning and a use that is different from those it had on the 'macro', in so far as here the potential of the initiation of emerging territorial scenarios for each VC is brought together with the ongoing local dynamics and governance. This analysis was directed in order to make potential actions emerge for the improvement of the ongoing phenomena. Thus TIE defines the position and role that each of the cities could play inside its own

12 Arco (Alto Garda-Ledro VC, 17,038 inhabitants), Pergine Valsugana (Alta Valsugana-Bersntol VC, 20,700 inhabitants), Riva del Garda (Alto Garda-Ledro VC, 16,052 inhabitants), Rovereto (Vallagarina VC, 38,141 inhabitants) and Trento (Val d'Adige VC, 115,540 inhabitants).

VC in respect to the three scenarios that were identified (Retail, Marketing and Landscape) (see Chapter 4).

The result of the application of TIE was the definition of a set of non-prescriptive strategies for territorial retail development, which were to be discussed and shared in the local decision-making process that each VC had to implement in 2015 in response to the innovative territorial planning tasks assigned to them by provincial institutional reform.

References

Alexander, E.R., 2005. Implementing Norms in Practice: The Institutional Design of Evaluation. In: D. Miller and D. Patassini (eds), *Beyond Benefit Cost Analysis*. Aldershot: Ashgate, pp. 295–310.

Alexander, E.R. (ed.), 2006a. *Evaluation in Planning: Evolution and Prospects*. Aldershot: Ashgate.

Alexander, E.R., 2006b. Evaluation and Rationalities: Reasoning with Values in Planning. In: E.R. Alexander (ed.), *Evaluation in Planning: Evolution and Prospects*. Aldershot: Ashgate, pp. 39–52.

Alexander, E.R., 2009. Dilemmas in evaluation planning, or back to basics: What is planning for? *Planning Theory & Practice*, 10: 233–44.

Archibugi, F., 2006. Pitfalls in Planning and Plan-evaluation. In: E.R. Alexander (ed.), *Evaluation in Planning: Evolution and Prospects*. Aldershot: Ashgate, pp. 73–83.

Archibugi, F., 2007. *Planning Theory: From the Political Debate to the Methodological Reconstruction*. Berlin: Springer.

Brunetta, G., 2006. Valutazione e pianificazione. Verso l'integrazione? *Scienze Regionali* [*Italian Journal of Regional Science*], 5: 119–26.

Brunetta, G., 2008. *Valutazione territoriale integrata degli insediamenti commerciali*. Firenze: Alinea.

Brunetta, G., 2013. Valutazione integrata territoriale per il governo del territorio [Territorial integrated evaluation in spatial planning]. *Scienze Regionali* [*Italian Journal of Regional Science*], 12: 71–91.

Brunetta, G. and Peano, A. (eds), 2003. *Valutazione ambientale strategica. Aspetti metodologici, procedurali e criticità*. Milan: IlSole24ore Edizioni.

Brunetta, G. and Voghera, A., 2008. Evaluating landscape for shared values: Tools, principles, and methods. *Landscape Research*, 33: 71–87.

Faludi, A. and Voogd, H. (eds), 1985. *Evaluation of Complex Policy Problems*. Delft: Delftsche Uitgers Maatschappij.

Guba, E.G. and Lincoln, Y.S., 1989, Putting It All Together so That It Spells E-V-A-L-U-A-T-I-O-N. In: E.G. Guba and Y.S. Lincoln. *Fourth Generation Evaluation*. Thousand Oaks, CA: SAGE Publications, pp. 252–69.

Hillier, J., 2011. Strategic planning as strategic navigation. *Town Planning Review*, 82: 503–27.

Khakee, A., 1998a. Evaluation and planning: Inseparable concepts. *Town Planning Review*, 69: 359–74.
Khakee, A., 1998b, The Communicative Turn in Planning and Evaluation. In: N. Lichfield, A. Barbanente, D. Borri, A. Khakee and A. Prat (eds), 1998. *Evaluation in Planning*. Dordrecht: Kluwer Academic Publishers, pp. 97–111.
Khakee, A., Hull, A., Miller, D. and Woltjer, J. (eds), 2008. *New Principles in Planning Evaluation*. Aldershot: Ashgate.
Lichfield, N., 1988. The Cost Benefit Approach in Plan Assessment. International Colloquium on Assessment Methods in Urban and Regional Planning: Theory and Case Study, Naples.
Lichfield, N., 1996. *Community Impact Evaluation*. London: University College Press.
Lichfield, N., 1998. Trends in Planning Evaluation: A British Perspective. In: N. Lichfield, A. Barbanente, D. Borri, A. Khakee and A. Prat (eds), 1998, *Evaluation in Planning*. Dordrecht: Kluwer Academic Publishers, pp. 1–15.
Lichfield, N., Barbanente, A., Borri, D., Khakee, A. and Prat, A. (eds), 1998. *Evaluation in Planning*. Dordrecht: Kluwer Academic Publishers.
Lichfield, N., Kettle, P. and Withbread, M., 1975. *Evaluation in Planning Process*. Oxford: Pergamon Press.
Lidblom, C.E., 1959. The science of muddling through. *Public Administration Review*, 39: 517–26.
Lidblom, C.E., 1979. Still muddling: Not yet through. *Public Administration Review*, 19: 79–88.
March, J.G. and Olsen, J.P., 1989. *Rediscovering Institution: The Organizational Basis of Politics*. New York: The Free Press.
March, J.G. and Olsen, J.P., 1995. *Democratic Governance*. New York: The Free Press.
Miller, D. and Patassini, D. (eds), 2005a. *Beyond Benefit Cost Analysis*. Aldershot: Ashgate.
Miller, D. and Patassini, D., 2005b. Introduction: Accounting for Non-Market Values in Planning Evaluation. In: D. Miller and D. Patassini (eds), *Beyond Benefit Cost Analysis*. Aldershot: Ashgate, pp. 1–11.
Moroni, S., 2006. Planning, Evaluation and the Public Interest. In: E.R. Alexander (ed.), *Evaluation in Planning*. Aldershot: Ashgate, pp. 55–71.
Palumbo, M., 2001. *Il processo di valutazione. Decidere, programmare, valutare*. Milan: Franco Angeli.
Patassini, D., 2006. Logiche valutative nei processi di pianificazione territoriale. In: S. Moroni and D. Patassini (eds), *Problemi valutativi nel governo del territorio e dell'ambiente*. Milan: Franco Angeli, pp. 23–66.
Voogd, H., 1998, The Comunicative Ideology and *ex ante* Planning Evaluation. In: N. Lichfield, A. Barbanente, D. Borri, A. Khakee and A. Prat (eds), *Evaluation in Planning*. Dordrech: Kluwer Academic Publishers, pp. 113–26.

Chapter 2
The Design of Territorial Integrated Evaluation[1]

Grazia Brunetta and Emma Salizzoni

2.1 The Rationale of the Multidimensional Evaluation Approach

Over the course of the twentieth century, theoretical and empirical research has approached the evaluation of retail urban policies mostly as a sector-focused question and has paid little attention to the territorial aspects connected with the processes of settlement (Wrigley, Coe and Currah, 2005). Especially in the Italian national context, the research (aside from few exemplary pioneering studies)[2] and the territorial governing practices have lacked any in-depth analyses of the territorial dynamics that retail activities put into motion. Rather, retail was mostly understood to be an object of study for economics.

The integration between *retail and 'territory'* is the constituting principle of the first National Reform Decree.[3] The decree takes an approach that is entirely different from the long-established treatment of retail policy as an issue involving only this sector. Instead, it shifts retail policy to a central position in the programming of territorial development. This reform introduced a new system of rules, constituted by a set of territorial criteria[4] and put regional spatial policies in charge of defining the conditions and strategies of territorial development (Boylaud and Nicoletti, 2001; Evans-Cowley, 2005). The renewed system of regulations and procedures for regional planning placed the territorial dimension in the centre of the process of reform. This territorial dimension became an element essential for addressing the dynamics of regional retail development. Several years later, this substantially innovative legislation was followed by the

1 This chapter is the result of the combined research activity undertaken by the two authors. The final written version of Chapter 2.1 can be attributed to Grazia Brunetta and that of Chapter 2.2 to Emma Salizzoni.

2 Such as Davies (1995),Hall (1988) and Sernini (1988).

3 See Law 114/1998, 'Reform of the retail sector discipline'.

4 The decree law gives up the long-established concepts of preceding legislation (zoning, maximum threshold, quantity). It attempts to begin a new planning approach through introducing new concepts like conurbation, mixité and retail diversification.

approval of a constitutional law[5] that gave regional administrations exclusive competence in territorial retail planning.

The territorial dimension is the fulcrum of this new approach. The effort lies in approaching the issues of development not as simple competitive relationships between retail sites but rather as complementary relationships among retail territorial systems (De Las Rivas and Morandi, 2007; Gibbs 2012; Guy, 2007; Thomas, Bromley and Tallon, 2003). Here retail is understood as an economic activity that can network with the other functions at work in a region. Despite this profound institutional innovation and the passing of 10 years since the reform process was activated, there are only a few Italian regions that have fully reached the objective of integration between retail and 'territory' in their regional development policies (Brunetta, 2007; Brunetta and Morandi, 2009).

As mentioned in the Introduction, the objective of taking the territorial dimension of retail into consideration is at the centre of the renewed framework of the regional development policies in the Autonomous Province of Trento (APT). The APT aimed to give life to a new course of Provincial policies that would harmonize the needs of territorial development with the needs for the conservation of landscape values.

TIE is a *multidimensional* process of meta-evaluation (see Chapter 1). It treats the evaluation of the territorial opportunities for retail development as an issue that is strictly interwoven with the evaluation of the dynamics of the landscape and the local economy. Thus TIE aims at designing scenarios where retail activities become a factor working to enhance not only the characteristics of the local economy but also of the territorial context and of the landscape. All this is an authentic *territorial* project; it is not a mere project for retail.

Because TIE is multidimensional, we had to take a multidisciplinary and interdisciplinary approach to evaluation. We had to engage in a constant dialogue, one that involved the disciplines that dealt with territorial planning, landscape planning, nature conservation, the conservation of historical and cultural resources, economic programming and territorial marketing. As TIE methodologies are being applied, these disciplines intersect with those related to surveying (see Chapter 4) and mathematics (see Chapter 5). These different fields of knowledge interacted, as did different conceptual and operational paradigms. This was the interaction that went through the entire development of our research as we collected the data, interpreted it in an integrated way, and defined the scenarios. Therefore we aimed to take up the challenges of a multidisciplinary and interdisciplinary process. This process is not only the sum total of various disciplines but their effective integration, a process that can be reflected, hopefully, in a corresponding multi- and inter-sector quality in the institutional context of the Provincial administrative units – the Province and the Valley Communities (VCs).

5 See Constitutional Act n. 3/2001 which reformed the Title V, Part Two of the Italian Constitution.

2.1.2 The Three Dimensions and Five Themes of the Evaluation

We identified five evaluation *themes* that fit within the perspective of the three closely interconnected *dimensions* of integrated evaluation (territorial, landscape and economic; see Chapter 1):

A. Territorial retail supply;
B. Tourism;
C. Land use;
D. Landscape;
E. Local governance.

Within this framework, the theme of retail (Theme A – 'Territorial retail supply') was interpreted as something to be, first of all, correlated to its *territorial dimension*. With regard to the territorial dimension, TIE focuses on the features of the demography, land use, and the infrastructure system (see Table 2.1). As far as demography is concerned, TIE attempts to detect the dynamics of the population in relation to potential local demand. In relation to land use, it attempts to outline the difference between territories that are extensively urbanized and territories prevalently dedicated to agriculture or grazing. Outlining these differences helps planners to identify the presence of productive sectors and/or urban services that could potentially be integrated with retail activity. In relation to the infrastructure system TIE used accessibility as an important criterion to help choose and define components of territorial retail scenarios in the Province (see Chapter 3.2 and Chapter 5). The reason is that retail activities, especially those related to the Gross Leasable Areas (GLAs), need to be supported by an infrastructure network that is effective, many-branched, and interconnected (Gruen and Smith, 1960; Kivell and Shaw, 1980). Besides all of these elements that can be included in Theme C – 'Land use' – non-material aspects of the territorial dimension were also analysed. Namely, Theme E – 'Local governance' – treats the level of initiative that characterizes the local territories of the Province when they design actions improving the quality of retail in city centres.

The integrated evaluation of material and non-material aspects of the territorial dimension allowed us to outline an overall picture of the territorial structure of the Province in its main physical and social components that provide potential support for the definition and activation of territorial retail scenarios.

Furthermore, TIE acknowledges the central position of landscape in the context of the Province as well as its increasingly significant role in urban and regional policies on a European and a national level (see the European Landscape Convention – ELC – proclaimed in 2000 and ratified in Italy in 2006). With this in mind, TIE is an effort to interpret and enhance the environmental and landscape dimension. It evaluates landscape in the Province territories (Theme D – 'Landscape') focusing on the values, critical issues and dynamics of natural and cultural resources. These resources are conceived not only as part of the

Table 2.1 **TIE themes and indicators identified on the micro and macro scales**

TIE themes	Micro scale	Macro scale
A. Territorial retail supply	A1m Retail consistency	A1M Retail consistency
	A2m Index of retail realization	A2M Retail operators
	A3m Retail typology	A3M Dynamics of retail operators
	A4m Retail specialization	A4M Dynamics of retail consistency
		A5M Retail density
		A6M Retail integration
		A7M Retail typology
		A8M Retail specialization
		A9M Index of retail choice
		A10M Index of retail loss
B. Tourism	B1m Arrivals	B1M Tourism intensity
	B2m Nights spent	B2M Tourism intensity (winter)
		B3M Tourism intensity (summer)
		B4M Dynamics of tourists
		B5M Accommodation capacity
		B6M Hospitality operators
		B7M Dynamics of hospitality operators
		B8M Extension of bicycle paths
C. Land use	C1m Urban morphology	C1M Dynamics of demography
	C2m Index of urban growth	C2M Urbanized areas
	C3m Agricultural areas	C3M Retail areas
	C4m Index of reusing no-longer-used areas and buildings	C4M Agricultural areas
	C5m Retail areas	C5M Pasture areas

	C6m Retail spatial distribution	C6M Infrastructure density
	C7m Retail spatial morphology	
D. Landscape	D1m Ecological network elements	D1M Ecological network elements
	D2m Geological and geomorphological elements, representative landscape areas	D2M High-quality agricultural areas
		D3M Geological and geomorphological elements, representative landscape areas
		D4M High-quality scenic views
		D5MD High-risk hydrogeological areas
		D6M High-danger hydrogeological areas
		D7M Traffic pollution
		D8M Air quality (PM10)
		D9M Index of urban growth
		D10M Index of loss of natural, semi-natural and agricultural land
E. Local governance	E1m Historical retail areas	E1M Municipal grants for the enhancement of historical retail areas
	E2m Historical shops	E2M Consortium grants for the enhancement of historical retail areas
	E3m Municipal grants for the enhancement of historical retail areas	
	E4m Consortium grants for the enhancement of historical retail areas	
	E5m Renewal projects for no-longer-used areas and buildings	
	E6m Events for the enhancement of retail and tourism	

Note: The indicators on the micro scale are marked by a small letter (m), the indicators on the macro scale are marked by a capital letter (M).
Source: Authors' elaboration.

common good but also as taking a cue from the orientation of the ELC,[6] as resources that can spur the development of the local economic system and, particularly, retail scenarios based on the integration of landscape resources, retail supply and the tourist system (see Chapter 3.2).

We should emphasize that TIE, in line with the ELC (sec. 5),[7] is a clear and meaningful example of the integration of landscape into sector-focused policies, here in relation to the programming and planning of the retail sector.

Usually, both in Italy and elsewhere in Europe, issues related to the landscape have had a hard time passing from the status of theoretical announcements,[8] which are widely agreed upon, to the status of practices that are effectively implemented through relevant policies (Brunetta and Voghera, 2008; Voghera, 2011). With specific reference to the national context, the landscape topics stipulated by the ELC have been recognized by the Code of Cultural Resources and Landscape (2004 and subsequent modifications and additions), though with several important gaps (Peano, 2011). Nevertheless, these topics have not yet been acted on in any widespread and effective way. In fact, there have been two evident and opposite prevalent tendencies. On the one hand, there is a passive safeguarding of the landscape heritage, bordering on its *museum-ification*; and, on the other hand, there is a total lack of attention towards landscape values and towards the risks that territorial development policies could bring (Lanzani, 2011; Paolinelli, 2011). In other words, the ELC (section 5d) had stipulated a road to take, but it has found few willing travellers. The ELC road is one that leads to the widespread care for *all* the landscape through policies of conservation, management and planning that are effectively integrated 'in general and sector-focused planning and programming on every institutional level' (Paolinelli, 2011: 52).[9]

6 'The landscape ... constitutes a resource favourable to economic activity and whose protection, management and planning can contribute to job creation' (European Landscape Convention, ELC, Preamble).

7 'Each Party undertakes ... to integrate landscape into its regional and town planning policies and in its cultural, environmental, agricultural, social and economic policies, as well as in any other policies with possible direct or indirect impact on landscape' (ELC, sec. 5).

8 Landscape has become a more and more current issue (Gambino, 2005) and has been recognized, under ELC influence, as a common resource and basis of the identity and the well-being of populations. Landscape, as a whole, should be the object of policies of conservation, management and planning (ELC, sec. 1, 5 6).

9 Motivations of 'opportunity' also support the integration of the landscape within territorial and sector-focused governmental policies. These go beyond the motivations of 'necessity' alluded to above – i.e. landscape as a common resource and basis for the identity of the population. The potential of landscape is recognized both as an indicator of sustainable development (Manzi, 2000; Molesti, 2008) and as a means to reach it. In fact, according to the ELC Preamble, landscape 'has an important public interest role in the cultural, ecological, environmental and social fields, and constitutes a resource favourable

This is the road that TIE strives to go down, finding aid and 'comfort' in the orientations of Provincial planning.[10] TIE integrates landscape into policies of retail planning both in the analytical-evaluative phase and the phase of designing scenarios (see Chapter 3.2). This kind of integration benefits both the landscape and the retail system. In fact, the landscape is protected from the potential impact of erroneously located or formatted GLAs, thanks to the formulation of sustainable retail scenarios. In addition, the retail system is potentially strengthened by the landscape, which is interpreted in its aspect as a resource for development. This is particularly true in those territories where landscape values are exceptionally high and where landscape can most effectively play its role as a resource for development in synergy with the retail and tourist systems.

Finally, TIE evaluates the dynamics of territorial economic activities and especially focuses its attention on tourist activity. One of its Theme, B – 'Tourism' – evaluates the intensity of tourist flows and the socioeconomic effect these flows have on the resident population. Theme B also evaluates the conditions of the hospitality facilities as well as of bicycle paths. In fact, tourism plays a leading role in the Provincial economy and can play a further role in correlating retail and landscape in an effective and sustainable way as new models of tourism are being defined in harmony with the exceptional natural and cultural resources.[11]

---

to economic activity'. Thus when planners respond to the aims of landscape quality, they can be, in fact, responding to the multiple aims of sustainable development effectively too. 'The fact is that by taking care of the landscape we simultaneously promote communal well-being, safeguard the environment and protect economic activity. All four ingredients of sustainable development (social, ecological, economic and cultural improvement) are thus involved here' (CoE, 2006, pp. 11–17).

10 Within the Italian context, the PTP is a particularly illuminated example of the integration of landscape into territorial governing policies according to the principles of the ELC (sec. 5). In fact, the Plan places landscape at the centre of its actions, attributing landscape values with 'the significance of inalienable sources of identity, inspirational criteria for planning on all levels, and essential cultural and economic resource' (PTP, exhibit A, p. 23). The plan does this with the specific purpose of bolstering the ability of Trento Province's territorial offerings to be recognized. In fact, the plan recognizes landscape identity as a factor that is important for competitiveness in tourist activities. In addition, the plan has a more general purpose of guaranteeing a sustainable development that should take place 'in profound coherence with the historical traditions and future aspirations of the inhabitants as well as with the environmental conformation of the territory and its overall carrying capacity' (PTP, exhibit A, p. 32).

11 In the Province the synergy between the landscape resources and the tourist system can become a means for moving scenarios for the distinction and improvement of retail in the following and in other ways: highly valued naturalistic landscape and excursion tourism. Broad areas of the Province are marked by landscape valued highly for its ecology and scenery. When the value of this landscape is emphasized appropriately, it constitutes an exceptional supply system, mainly for a type of excursion- and/or sports-related tourism. This type of tourism can help the Valley Communities (VCs) to 'cross over' towards general objectives of greater sustainability. In the perspective of reducing the

The aim is to help develop policies that integrate natural resources, the tourist industry, and local agricultural activities. Such an integrated system can put forward offerings that are rooted in the territorial context, attractive, and firmly anchored in the local characteristics.[12]

2.2 The Evaluation Matrix

Following the multidimensional approach described above, we designed an evaluation matrix characterized by the five TIE themes and the indicators related to them (Table 2.1). These indicators were identified in reference to two territorial scales of evaluation, the micro scale (indicated by a small 'm') and the macro scale (indicated by a capital 'M'): the macro scale, covering the 16 Valley Communities; and the micro scale, covering the five most populous cities of the Province (Trento, Rovereto, Riva del Garda, Pergine Valsugana and Arco).

The application of the matrix of indicators to the Province territories allowed us to highlight factors of strength and weakness as well as opportunity and threat in relation to the five evaluation themes, respectively on a micro and macro scale (see chapters 1 and 3). The results of this evaluation became the basis for the definition of territorial retail scenarios. Therefore, when we went through the

---

seasonal nature of the tourist flows, the VCs could diversify their offerings, going beyond ski tourism, which is dominant. In addition, they could advance towards territorial retail scenarios characterized by a supply system that is appropriate to its territorial context, which is highly specialized and founded on formats and structures oriented towards the preservation of current landscape values: rural landscape and eno-gastronomical tourism. Although rural landscape makes up only 10 per cent of the Province surface, it plays a leading role in the context of Trento Province. It is a precious resource, especially in those territories where there are high numbers of 'High-quality agricultural areas' (PTP, exhibit A: 64). In these cases, the enhancement of the landscape heritage for tourism is the main conduit for guaranteeing the development of retail scenarios founded on the local characteristics, and especially on the promotion of local products – DOP (origin-guaranteed foods) and DOC (origin-guaranteed wines) – according to appropriate retail formats and structures: historical and identity landscape and cultural tourism. The landscape of Trento Province has a high incidence of resources valued for their historical importance or importance for local identity. These are not only resources that are now being brought to the light, but also those that are already in place in widespread 'systems' of resources. Such resources can be enhanced effectively and become systems for public access to culture – for example, through planners helping to promote them organically and making sure that they can be accessed. If this happens, resources could provide the drive for a prevalent cultural tourism that, especially in an urban context, could give an important lift to the development of scenarios founded on the retail enhancement of historical centres, scenarios addressing formats and structures appropriate for already established city dimensions.

12 These aims were introduced in the last Provincial Programme Development for the fifteenth legislature, approved by the Provincial Council in December 2014.

process of choosing and defining the indicators for each theme and consequently formulated the matrix, we took a decisive step in developing our research. In fact, this process enabled us to conduct an effective evaluative analysis in order to define the scenarios.

Before we begin to illustrate the characteristics of the indicators identified for the five TIE themes, it is important to state that the process of formulation of the matrix was conducted in the light of the prospective for territorial renewal, enhancement and development outlined in the PTP.[13] From the perspective of TIE's triple evaluation dimension – not only economic but also territorial and landscape-related – our intent was to anchor the TIE evaluation firmly to the principles and strategic guidelines in the PTP.

The PTP is based on two kinds of principles – conditions and orientations. First, conditions consist in the so-called 'invariants'.[14] The invariants are directly regulated by the PTP[15] and are immediately prescriptive. Thus they needed to be taken into consideration by our research. In fact, they were very influential in the analytical-evaluative phase of TIE.[16] Second, orientations consist in objectives that are not immediately prescriptive. These objectives, that were directly correlated with TIE's leitmotif and aim (the definition of territorial retail scenarios), were extrapolated as points of reference for the definition of the evaluation matrix. Moreover, the PTP defines four strategic guidelines, which should be interpreted as its 'structural recommendations' (PTP, exhibit E, p. 5):

---

13 The indicators were also defined in view of the regulatory framework of the Province in reference to the conduct of retail activity, especially the framework of Provincial Law 17/2010 in reference to criteria for the settlement of new GLAs: containment of the territorial and environmental impact of large sized structures, promotion of the territorial quality, of the urban texture and of the historical centres in addition to the protection of the environment from pollution and protection of health (sec. 11.2).

14 That is, 'the territorial elements that constitute the distinctive characteristics of the territorial environment and identity, in as much as they are stable configurations or ones whose modifications proceed slowly and they deserve to be protected and enhanced in order to guarantee balanced and sustainable development in the evolutionary processes that are foreseen and promoted by the instruments of territorial planning' (PTP, sec. 8). The PTP considers the following elements as 'invariants': the resources of the Dolomite Mountains heritage; the hydrographic network (rivers, lakes and the glaciers); forests and high natural areas; high-quality agricultural areas; the geological and geomorphological elements; the representative landscapes (environmental resources, representative archaeological resources, and representative architectural and artistic resources).

15 'The PTP is structured as an instrument of territorial coordination and of the regulation of the invariants or the permanent components of the territory …' (PTP, exhibit A, p. 89).

16 Several indicators were defined ad hoc, that is D1m, D2m, D1M, D2M, D3M (see Table 2.1, Table 2.3 and Table 2.4).

> *identity* (reinforcing the recognition level of the territorial offerings in Trento Province, emphasizing the worth of the varied landscape, the environment quality and individual cultural characteristics), *sustainability* (promoting sustainable development through limiting soil sealing and the consumption of primary resources as well as through privileging urban and territorial renewal), *integration* (consolidating the integration of Trento Province into the European context through effectively linking it into the great networks of infrastructure, the environment, the economy, the society, and culture) and *competitiveness* (reinforcing local capacities for self-determination and competitiveness as well as the opportunities for the lasting development of the whole provincial system). (PTP, exhibit E, p. 5)

Table 2.2 following is a synoptic outline tracing the direct correlation and consistency between the content of the PTP – its principles and strategic guidelines – and some of the indicators of the TIE matrix, defined on the micro and on the macro scale.

2.2.1 The Indicators

The TIE evaluation matrix is composed of 21 indicators on the micro scale and 36 indicators on the macro scale (see Table 2.2, Table 2.3, Table 2.4). Some of the indicators are common to both scales, functioning in a complementary and synergetic way for the analysis and evaluation of the five themes.[17]

The principal objective that guided us when we were choosing and defining the indicators that are the components of the matrix was that we wanted to achieve *a complex and dynamic representation* of the themes that were the objects of analysis and evaluation, detecting both their qualitative and qualitative aspects as well as their evolution over time. An illustrative example of this consists in the indicators we chose to evaluate: Themes A and D, 'Territorial retail supply' (A) and 'Landscape' (D). The indicators for Theme A reveal the quantitative consistency of the retail phenomenon, but do not stop at that. They reveal also the characteristics of the types of retail (A6M – 'Retail integration', A7M – 'Format typology' and A8M – 'Retail specialization'). In addition, they trace the evolution of the territorial retail system over the recent years (A3M – 'Dynamics of retail operators', A4M 'Dynamics of retail consistency'). Similarly the indicators for Theme D, 'Ecosystem and Landscape', attempt to

---

17 Indicators A1m/A1M – 'Retail consistency', A3m/A7M – 'Format typology', A4m/A8M – 'Retail specialization, C2m/D9M 'Index of urban growth, C3m/C4M – 'Agricultural areas', C5m/C3M – 'Retail areas', D1m/D1M – 'Ecological network elements', D2m/D3M – 'Geological and geomorphological elements, representative landscape areas', E3m/E1M – 'Municipal grants for the Enhancement of Historical Retail Areas', E4m/E2M – 'Consortium grants for the Enhancement of Historical Retail Areas'.

get the intrinsically complex and dynamic characteristics of the landscape.[18] In fact, they manage to delineate the plurality of landscape values when they describe their ecological (D1M – 'Ecological network elements'),[19] scenic (D4M – 'High-quality scenic views') and cultural-historical and identity-related values (D3M – 'Geological and geomorphological elements, representative landscape areas').[20] In addition, they trace the main processes that transform these values (D9M – Index of urban growth' and D10M 'Index of loss of natural, semi-natural and agricultural areas').

At the same time, we were careful to limit the overall number of indicators. We worked out *an evaluation grid* that was substantially *complete but* that was *also agile* and easily to work with. We stayed close to the methodology of TIE, which aims to act as an institutional learning tool that supports the processes of territorial governance (see Chapter 1). Therefore, we performed the technical operation of designing the system of indicators for evaluating and simulating territorial retail scenarios in a way that they could be easily applied. (The number of indicators was limited. Validated data could be retrieved easily and periodically updated by various agencies in public administration). In this way, the five evaluation themes could become operative for actions of designing and monitoring the emerging scenarios of enhancement and development in the Province.

Another objective that guided us while we were structuring the evaluation matrix was that we wanted to identify *indicators* that were *correlated to the particular characteristics of the territorial, landscape-related and economic system* in the Province, and therefore those rooted in the local conditions. For example, the indicators for Theme E, particularly those on the micro scale, specifically refer to the normative initiatives of Provincial laws (in fact, the following indicators all refer to the retail Provincial regulations, Provincial Law

18 Landscape is complex because 'the landscape has an important public interest role in the cultural, ecological, environmental and social fields, and constitutes a resource favourable to economic activity …' (ELC, Preamble). Landscape is dynamic because 'it inevitably changes and evolves over time, in response to natural processes and to the changing needs and activities of people' (Phillips, 2005: 21).

19 The elements that make up the ecological network of the Province are the following: (1) the resources of the Dolomite Mountains heritage; (2) the hydrographic network (rivers, lakes and the glaciers); (3) forests and high natural areas (state-owned forests, the Natura 2000 areas, Provincial and National Parks and Provincial Natural Reserves); and (4) high-quality agricultural areas. These elements are identified by the PTP (sec. 8) as 'invariants'.

20 The following elements were considered: (1) the geological and geomorphological elements; (2) the representative landscapes (environmental resources, representative archaeological resources, and representative architectural and artistic resources). These elements are identified by the PTP (sec. 8) as 'invariants'.

Table 2.2 The relationship between TIE indicators and the principles and strategic guidelines of the PTP

PTP Principles	*PTP Strategic guidelines*			
	Identity	**Sustainability**	**Integration**	**Competitiveness**
I. Conditions				
Invariants	D2M High-quality agricultural areas			
	D2m/D3M Geological and geomorphological elements, representative landscape areas	D1m/D1M Ecological network elements		
	D4M High-quality scenic views			
II. Orientations				
Limitation of soil sealing	D9M Index of urban growth	D9M Index of urban growth		
	D10M Index of loss of natural, semi-natural and agricultural land	D10M Index of loss of natural, semi-natural and agricultural land		
Reuse of no-longer-used buildings		C4m Index of reusing no-longer-used areas and buildings		

		E5m Renewal projects for no-longer-used areas and buildings	
Renewal and enhancement of historical city centres	E1m Historical retail areas		E1M Municipal grants for the enhancement of historical retail areas
	E2m Historical shops		E2M Consortium grants for the enhancement of historical retail areas
Integration of retail with other sectors of the economy			E6m Events for the enhancement of retail and tourism
Integration of retail with linking infrastructures and complementary services			B8M Extension of bicycle paths

Source: Authors' elaboration.

17/2010, sec. 63 and 64: E1m – 'Historical retail areas',[21] E2m – 'Historical shops',[22] E3m – 'Municipal grants for the enhancement of historical retail areas', and E4m – 'Consortium grants for the enhancement of historical retail areas').[23] Also, the indicators for Theme B, Tourism, were calibrated to the characteristics of the local phenomena. For example we identified several indicators that reveal the tourist flows in the winter and summer seasons, distinguishing between them, that is, B2M – 'Tourism intensity (winter)' and B3M – 'Tourism intensity (summer)'. The territories of the Province are endowed with such geomorphological variety that they are subject to substantially varied and distinct kinds of tourist demand territorially according to the offerings (e.g. ski facilities or swimming facilities) and hence according to the season (winter or summer). This is the demand that we strove to trace through these indicators in an opportune way. Similarly, we added the indicator B8M – 'Consistency of Bicycle Paths', among those related to the Theme 'Tourism', because of the leading role that bicycle paths play and the flows of tourists that these attract within the overall Provincial tourist system.

Finally, we wanted to choose *indicators on the basis of which evaluation scale they would be applied to*. We chose indicators that more appropriately described the various phenomena on the micro (urban) scale and indicators more appropriate for the macro (VC) scale. For example, with regard to the Theme C – 'Land use' – we defined several indicators on the micro scale in order to collect the spatial conformation of urban phenomena (C1m – 'Urban morphology') and that of retail phenomena (C6m – 'Retail spatial distribution').[24]

21 'Historical retail areas' are defined as 'those portions of cities where there is the consolidated and widespread presence of retail settlements, understood as those areas that traditionally and historically have been devoted to the acquisition and trade of merchandise and objects' (Provincial Law 539/2011, exhibit A, sec. 1).

22 'Historical shops' are retail shops that 'have been in business in the same places with the same type of merchandise (or similar types) at least for 50 years and where the setting of the shop in its locale or its decoration is characterized by elements, instruments, tools and documents of special historical, artistic, architectural, environment and cultural interest or elements that are particularly significant for the language and culture of the place' (Provincial Law 539/2011, exhibit B, sec. 2).

23 Provincial Law 17/2010 lays down the possibility for cities and private actors (Consortiums) to receive Provincial contributions destined for the producing initiatives that enhance the historical retail areas (addressed to cities, sec. 64.1) or destined for the organization of events with appeal for retail and tourism in such areas (addressed to consortiums, sec. 64.2). Not all the cities and consortiums in Trento Province can benefit from such contributions because the Province identified specific requirements that cities and consortiums must have in order to be able to present requests for funding. (See Resolution 1339/2013, 448/2011 for the cities, and Resolution 844/2010 and 2868/2010 for the consortiums.)

24 In order to construct a territorial representation of the distribution structure of the retail supply, we identified and classified the morphologies of the retail settlements –

These are phenomena that cannot be detected on a territorial (macro) scale. Also with reference to the Theme E – 'Local Governance' – we chose micro scale indicators as they were able to describe projects and initiatives developed on a local scale and that would be hard to detect on a territorial (macro) scale (that are E5m – 'Renewal projects for no-longer-used areas and buildings' and E6m – 'Events for the enhancement of retail and tourism').[25]

We should now briefly outline the methodological process that, based on the calculation of the indicators, brought us to evaluate the overall performance of the themes for each territorial context.

Starting from the available data validated and updated periodically in the statistical archives of various administrative departments of the Province,[26] the indicators were calculated primarily on the basis of the logical-mathematical structure. Then, the emerged results were evaluated in reference to a benchmark, which is defined as the average value of the indicators in the context of reference. In the case of the VCs, the context of reference is the whole Trento Province. In the case of the cities, the context of reference is the total of the five cities taken as a whole. Therefore each indicator can take on a value more or less than the average[27] in the specific territorial context, VC or City (See Table 2.5).

---

that is, those contexts where various retail typologies are concentrated. These form spatial configurations characterized by their types of settlements and offerings, marked by a recognized and consolidated structures: (1) *linear agglomerations* – groups of retail structures located along the main traffic roads inside the cities and between cities; (2) *concentrated agglomerations* – groups of retail structures located near crossroads of main traffic infrastructure or crossroads in consolidated city centres (historical centres and suburban areas); and (3) *single settlements* – single large retail structures in suburban areas.

25 Initiatives for the integrated enhancement of retail and tourism are defined as events that correlate local resources, tourist flows and retail supply (typically, fairs, farmers markets and cultural and eno-gastronomic events). These are events that are located in urban areas (mainly in the historical city centres) and characterized by being held annually.

26 The multidimensional nature of the TIE evaluation process made it necessary for us to interact with various sectors of the public administration that handled, for example, the fields of economic development, territorial planning, and natural and cultural resource conservation. The TIE process was not helped by the scarcity of inter-sector contacts in public administrations. This made it necessary for us to encounter a variety of individuals and use a variety of ways of collecting and classifying data. Because of this, we initiated a process whereby we cross-checked the data and standardized it.

27 We chose to relate our calculation of the indicators to the values of the mathematical average taken in our context. We did not calculate it in relation to the corresponding values of the median. This choice was motivated by the fact that the mathematical average helps us more clearly characterize the individual territorial contexts in relation to various indicators and therefore in relation to various topics. In fact, from the point of view of statistics, the average is strongly sensitive to the changes made at the

Table 2.3 Description of the indicators on the micro scale

	Name	Description
A. Territorial retail supply	A1m Retail consistency	City retail surface/VC retail surface
	A2m Index of retail realization	City realized GLAs/VC realized GLAs
	A3m Retail typology	City neighbourhoods shops/VC total shops
	A4m Retail specialization	City food and non-food GLAs/VC GLAs
B. Tourism	B1m Arrivals	City number of arrivals/VC number of arrivals
	B2m Nights spent	City number of nights spent/VC number of nights spent
C. Land use	C1m Urban morphology	Classification of urban morphology (concentrated or sprawled)
	C2m Index of urban growth	City surface of urbanized ex-natural, semi-natural, and agricultural areas/city surface of urbanized areas
	C3m Agricultural areas	City agricultural surface/city territorial surface
	C4m Index of reusing no-longer-used areas and buildings	City reused no-longer-used surface/city industrial surface
	C5m Retail areas	City retail surface/city territorial surface
	C6m Retail spatial distribution	City number of retail agglomeration
	C7m Retail spatial morphology	Food and non-food GLAs/GLAs in every retail agglomeration
D. Landscape	D1m Ecological network elements	City surface of the ecological network elements/city territorial surface
	D2m Geological and geomorphological elements, representative landscape areas	City number of geological and geomorphological elements, representative landscape areas/VC number of geological and geomorphological elements, representative landscape areas
E. Local governance	E1m Historical retail areas	Presence or absence of recognized historical retail areas
	E2m Historical shops	Presence or absence of retail shops enrolled in the Official Register of Historical Shops
	E3m Municipal grants for the enhancement of historical retail areas	Presence or absence of municipal grants
	E4m Consortium grants for the enhancement of historical retail areas	Presence or absence of consortiums grants
	E5m Renewal projects for no-longer-used areas and buildings	Number of renewal projects for no-longer-used areas and buildings
	E6m Events for the enhancement of retail and tourism	Number of events for the enhancement of retail and tourism

Source: Authors' elaboration.

Table 2.4 Description of the indicators on the macro scale

	Name	Description
A. Territorial retail supply	A1M Retail consistency	VC retail surface/Province retail surface
	A2M Retail operators	VC retail operators/VC total operators
	A3M Dynamics of retail operators	VC variation in retail operators (2001–2009)
	A4M Dynamics of retail consistency	VC variation in retail surface (2005–2010)
	A5M Retail density	VC number of shops for every 1,000 inhabitants
	A6M Retail integration	VC retail surface by merchandise category (food and non-food)/VC retail surface
	A7M Retail typology	VC neighbourhoods shops/Province total shops
	A8M Retail specialization	VC Food and non-food GLAs/Province GLAs
	A9M Index of retail choice	VC number of GLAs within a 15-minute drive
	A10M Index of retail loss	Number of GLAs that intercepted the VC within 90-minute drive
B. Tourism	B1M Tourism intensity	VC annual nights spent/VC inhabitants
	B2M Tourism intensity (winter)	VC winter nights spent/VC inhabitants
	B3M Tourism intensity (summer)	VC summer nights spent/VC inhabitants
	B4M Dynamics of tourists	VC variation in annual nights spent (2001–2010)
	B5M Accommodation capacity	VC number of hospitality facilities/VC inhabitants
	B6M Hospitality operators	VC number of hospitality operators (hotels, restaurants)/VC number of operators
	B7M Dynamics of hospitality operators	VC variation in the number of hospitality operators (hotels, restaurants) 2001–2009
	B8M Extension of bicycle paths	VC extension of existing and foreseen bicycle paths/extension of Province existing and foreseen bicycle paths
C. Land use	C1M Dynamics of demography	VC variation in resident population (2001–2010)
	C2M Urbanized areas	VC urbanized surface/VC territorial surface
	C3M Retail areas	VC retail surface/VC territorial surface
	C4M Agricultural areas	VC agricultural surface/VC territorial surface
	C5M Pasture areas	VC pastured surface/VC territorial surface
	C6M Infrastructure density	VC infrastructure surface/VC territorial surface

Table 2.4 continued

	Name	Description
D. Landscape	D1M Ecological network elements	VC surface of the ecological network elements/VC territorial surface
	D2M High-quality agricultural areas	VC surface of high-quality agricultural areas/VC territorial surface
	D3M Geological and geomorphological elements, representative landscape areas	VC number of geological and geomorphological elements, representative landscape areas/Province number of geological and geomorphological elements, representative landscape areas
	D4M High-quality scenic views	VC number of high-quality scenic views/Province number of high-quality scenic views
	D5MD High-risk hydrogeological areas	VC surface subject to high hydrogeological risk/VC territorial surface
	D6M High-danger hydrogeological areas	VC surface subject to high hydrogeological danger (river flooding, landslides, avalanches)/VC territorial Surface
	D7M Traffic pollution	VC annual emissions of macro-pollutants from traffic (tons of NOx, PM10, COV, CO)
	D8M Air quality (PM10)	VC annual concentration of PM10/max permitted concentration (40 mcg/mc, Decree 155/2010)
	D9M Index of urban growth	VC surface of urbanized ex-natural, semi-natural and agricultural areas/VC surface of urbanized areas
	D10M Index of loss of natural, semi-natural and agricultural land	VC surface of urbanized ex-natural, semi-natural and agricultural areas/VC surface of natural, semi-natural and agricultural areas
E. Local governance	E1M Municipal grants for the enhancement of historical retail areas	Presence or absence of municipal grants
	E2M Consortium grants for the enhancement of historical retail areas	Presence or absence of consortiums grants

Source: Authors' elaboration.

Table 2.5 Calculation of the indicator A1M – 'Retail consistency'

Description	Criterion			Definition
		VC 12	0.9%	Low retail consistency
		VC 05	0.9%	
		VC 16	0.9%	
		VC 14	1.0%	
		VC 02	1.7%	
	Min consistency VC = 0.9%	VC 11	3.1%	
		VC 07	3.3%	
		VC 01	4.0%	
VC retail surface/ Province retail surface	Average consistency VC = 6.3%	VC 03	4.5%	
	Max consistency VC = 23.9%	VC 06	6.4%	High retail consistency
		VC 13	6.6%	
		VC 08	7.5%	
		VC 04	7.5%	
		VC 09	10.1%	
		VC 10	17.6%	
		VC 15	23.9%	

Source: Authors' elaboration.

Through the application of the SWOT analysis, the values obtained were translated into strength or weakness factors for the cities and opportunity or threat factors for the VCs.[28] This operation led to an overall evaluation of each themes.[29]

Such an overall process of calculation and evaluation of the indicators was supported by the elaboration of geo-referring maps that highlight the distribution and the spatial dimension of the phenomena analysed. In particular, the maps represented the structure of the retail supply and the territorial and landscape use (Themes A, C and D of the TIE matrix) (see Chapter 3).

extremes of a distribution of data (minimums and maximums). In contrast, the median, which is structured as an index of position, is not sensitive to these changes.

28 For example, the high value on the indicator D9M – 'Tendency towards the expansion of urbanized areas' – was interpreted as a threatening factor for the VC.

29 The overall evaluation of each theme was done by considering which indicators that the theme consisted in occurred more – factors of weakness/threat or factors of strength/opportunity. For example, if in a VC five indicators over eight on the theme 'Tourism' demonstrated values above the averages in the context of reference (Province), the overall theme, '*Tourism*', was considered as making up a factor of opportunity for the VC.

References

Boylaud, O. and Nicoletti, G., 2001. Regulatory reform in retail distribution. *Ocse Economic Studies*, 32(1): 253–74.

Brunetta, G., 2007. Geografia delle localizzazioni commerciali e sistemi di regolazione e valutazione in Piemonte. *Ciudades: Revista anual del Istituto universitario de Urbanistica de la Universitad de Valladolid*, 10: 97–108.

Brunetta, G. and Morandi, C. (eds), 2009. *Polarità commerciali e trasformazioni territoriali. Un approccio interregionale*. Firenze: Alinea.

Brunetta, G. and Voghera A., 2008. Evaluating landscape for shared values: Tools, principles, and methods. *Landscape Research*, 33(1): 71–87.

Council of Europe (CoE), 2000. *European Landscape Convention*.

Council of Europe (CoE), 2006. *Landscape and Sustainable Development: Challenges of the European Landscape Convention*.

Davies, R.L. (ed.), 1995. *Retail Planning Policies in Western Europe*. London: Routledge.

De Las Rivas, J.L. and Morandi, C. (eds), 2007. Generadores de nueva urbanidad: Los espacios comerciales. *Ciudades: Revista anual del Istituto universitario de Urbanistica de la Universitad de Valladolid*, 10: 19–240.

Evans-Cowley, J.S., 2005. Regulatory strategies for big-boxes. *Zoning Practice*, 22(12): 1–7.

Gambino, R., 2005. La gestione del paesaggio nelle aree protette. In: M. Sargolini (ed.), *Paesaggio, territorio del dialogo*. Rome: Edizioni Kappa.

Gibbs, R.J., 2012. *Principles of Urban Retail Planning and Development*. Hoboken, NJ: Wiley.

Gruen, V. and Smith, L., 1960. *Shopping Town USA: The Planning of Shopping Centers*, New York: Reinhold.

Guy, C., 2007. *Planning for Retail Development*. Abingdon: Routledge.

Hall, P., 1988, Planning for retail change: International comparisons. *Built Environment*, 14(1): 5–6.

Kivell, P.T. and Shaw, G., 1980. The Study of Retail Location. In: J.A. Dawson (ed.), *Retail Geography*. New York: Wiley, pp. 95–155.

Lanzani, A., 2011. *In cammino nel paesaggio. Questioni di geografia e urbanistica*. Rome: Carocci Editore.

Manzi, E., 2000. Mediterranean Concentration and Landscape: Six Cases. In: R. King, P. de Mas and J. Mansvelt Beck (eds), *Geography, Environment and Development in the Mediterranean*. Brighton-Portland: Sussex Academic Press.

Molesti, R., 2008. Il paesaggio quale sensore dello sviluppo sostenibile. Il capitale naturale e la sopravvivenza umana – Le trasformazioni del paesaggio come testimonianze dello sviluppo economico, *Economia e Ambiente*, 1–2: 2–4.

Paolinelli, G., (ed.), 2011. *Habitare. Il paesaggio nei piani territoriali*. Milan: Franco Angeli.

Peano, A., 2011. In favore di un unico piano paesaggistico e territoriale. In: A. Peano (ed.), *Fare paesaggio. Dalla pianificazione di area vasta all'operatività locale*. Firenze: Alinea Editrice.

Phillips, A., 2005. Landscape as a Meeting Ground: Category V Protected Landscapes/Seascapes and World Heritage Cultural Landscapes. In: J. Brown, N. Mitchell and M. Beresford (eds), *The Protected Landscape Approach: Linking Nature, Culture and Community*. Gland: IUCN.

Sernini, M., 1988. I centri commerciali integrati in Italia. Quando il developer diventa urbanista. *Archivio di Studi Urbani e Regionali*, 33: 3–28.

Thomas, C.J., Bromley, R.D.F. and Tallon, A.R., 2003. Retail park revisited: A growing competitive threat to traditional shopping centres? *Environment and Planning A*, 36: 647–66.

Voghera, A., 2011. *Dopo la Convenzione Europea del paesaggio. Politiche, piani e valutazione*. Firenze: Alinea Editrice.

Wrigley, N., Coe, N. and Currah, A., 2005. Globalizing retail: Conceptualizing the distribution-based transnational corporation (TNC). *Progress in Human Geography*, 29(4): 437–57.

Chapter 3
The Experimentation of Territorial Integrated Evaluation[1]

Ombretta Caldarice and Emma Salizzoni

3.1 Analysis of the Spatial Dynamics

This chapter reports the results that came out of the application of the Territorial Integrated Evaluation (TIE) matrix (see Chapter 1). The analysis developed at the macro scale gathers the performances of each Valley Community (VC) in reference to the Provincial context. The analysis conducted at the micro scale examines in depth the dynamics evidenced on the macro scale in the five most populous cities in the Trentino (Arco, Pergine Valsugana, Riva del Garda, Rovereto, Trento).

3.1.1 Territorial Retail Supply: A Complex and Varied Structure

On the macro scale, the analysis of Theme A – 'Territorial retail supply' – points out that the territories of Trentino are marked by retail systems that are very different from each other. On the one hand, the five most populous VCs in the Province, located along the main roads (Val d'Adige, Rotaliana-Könisberg, Vallagarina, Alto Garda-Ledro and Alta Valsugana-Bersntol VCs), are characterized by solid retail systems. Quantitatively, these VCs are marked by high retail consistency (A1M, see Chapter 2.2) and a large number of gross leasable areas (GLAs) and mid-sized retail shops (MRSs).[2] Qualitatively, they are marked by a high level of retail specialization (A8M). Hence their retail system is a factor of opportunity for the development of these territories. Moreover, these VCs are characterized by a high index of retail choice (A9M),

1 The chapter is the result of the combined work of the two authors. The final written version of Chapter 3.1 can be attributed to Ombretta Caldarice and that of Chapter 3.2 to Emma Salizzoni.

2 Mid-sized retail shops (MRSs) are retail stores whose sales areas are more than 150 sq m and up to 400 sq m in cities with 5,000–10,000 permanent inhabitants and up to 800 sq m in other cities. GLAs are retail stores whose sales areas are more than 300 sq m in cities with fewer than 5,000 inhabitants, more than 400 sq m in cities with 5,000–10,000 permanent inhabitants, and more than 800 sq m in other cities (Provincial Law 17/2010, sec. 3).

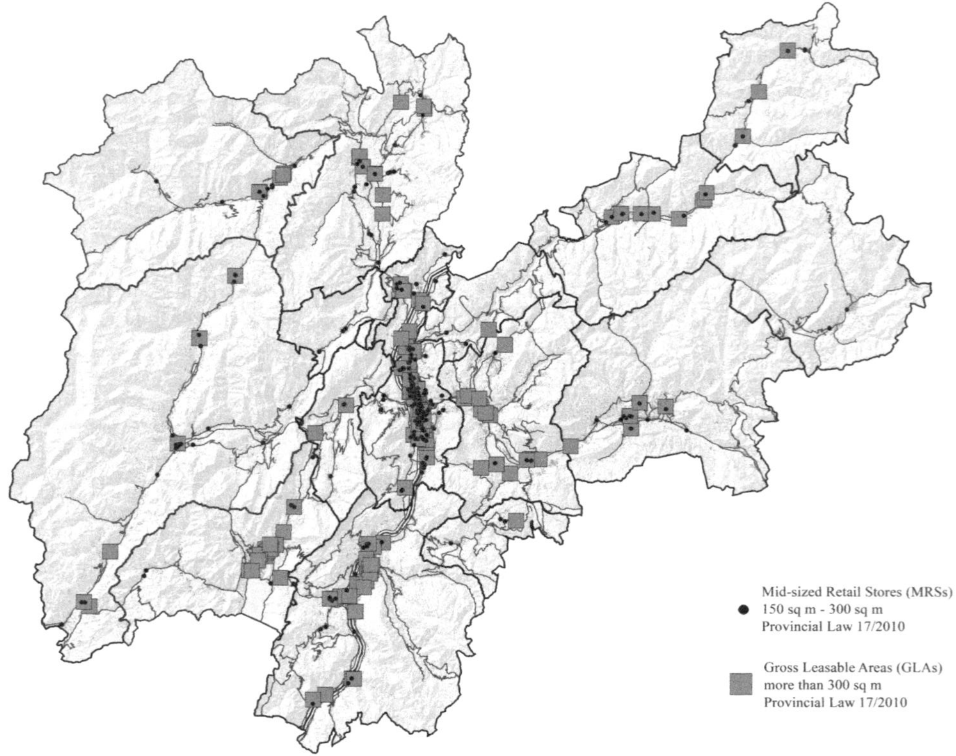

Map 3.1 The retail system in Trentino

Source: Authors' elaboration based on data elaborated by the Department of Economic Development and Work of the APT. All subsequent figures are from the same source, unless otherwise stated.

Map 3.2 Territorial retail supply: Opportunity and Threat factors in the VCs

Note: In dark grey the opportunity factors, in light grey the threat factors.

thanks to the presence of GLAs, and, at the same time, by a high index of retail loss (A10M).[3] The first reason for this is that they are located on the limits of the Province, as in the case of Alto Garda-Ledro and Vallagarina VCs. Hence they are affected by the influence of the competing retail areas in Lombardia and Veneto. The second reason is that they have a well-developed transportation infrastructure, as in the case of Rotaliana-Könisberg and Val d'Adige VCs which are crossed by the north–south Brenner motorway that links Veneto and the Bolzano Province.

3 The index of retail choice (A9M) is the number of market areas of the GLAs and shopping centres that can be reached by car within 15 minutes. The index of retail loss (A10M) is the number of competing market areas of the GLAs outside the Trento Province that can be reached within 90 minutes.

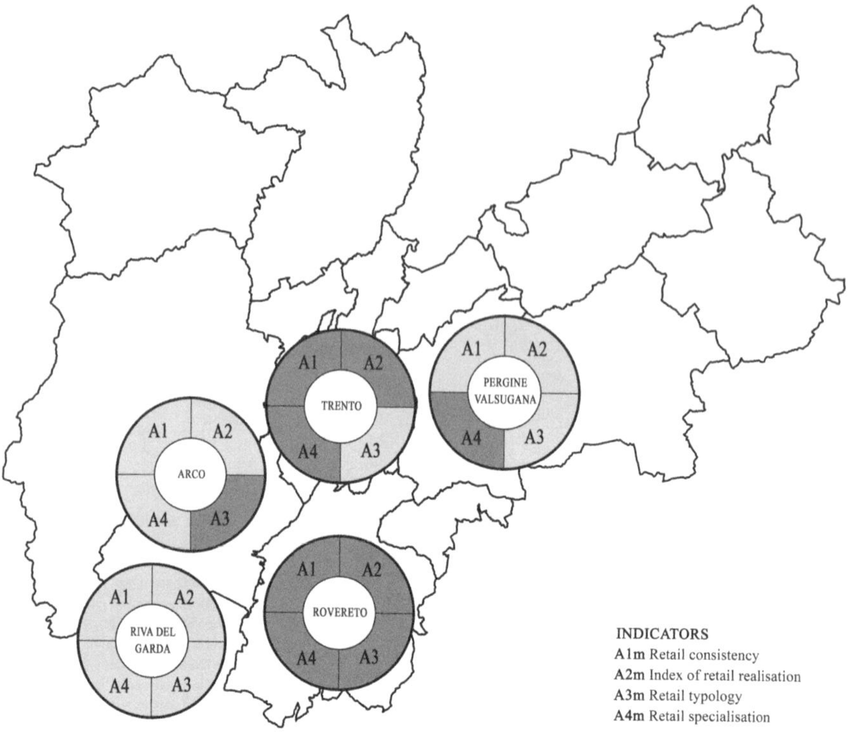

Map 3.3 Territorial retail supply: Strength and Weakness factors in the cities

Note: In dark grey the strength factors, in light grey the weakness factors.

On the other hand, the rest of the VCs in Trentino are marked by low retail consistency (A1M), yet they have a higher number of neighbourhood stores[4] than MRSs and GLAs. Furthermore, some of these VCs are marked by a good level of retail specialization (A8M) and low index of retail choice (A9M) and retail loss (A10M).

On the micro scale, the five cities in Trentino with over 10,000 inhabitants are located in the VCs with a good system of retail supply (alluded to above). In fact, Trento, Rovereto, Arco, Riva del Garda and Pergine Valsugana are located respectively in Val d'Adige, Vallagarina, Alto Garda-Ledro and Alta Valsugana-Bersntol VCs. Among these five cities, Trento and Rovereto are outstanding because their retail systems are strong points in the local economy, being high in consistency (A1m), well specialized (A4m), and highly varied in their typologies (A3m). On the other hand, Riva del Garda has a weaker retail context marked

4 Neighbourhood stores are retail stores whose sales area is not more than 150 sq m (Provincial Law 17/2010, sec. 3).

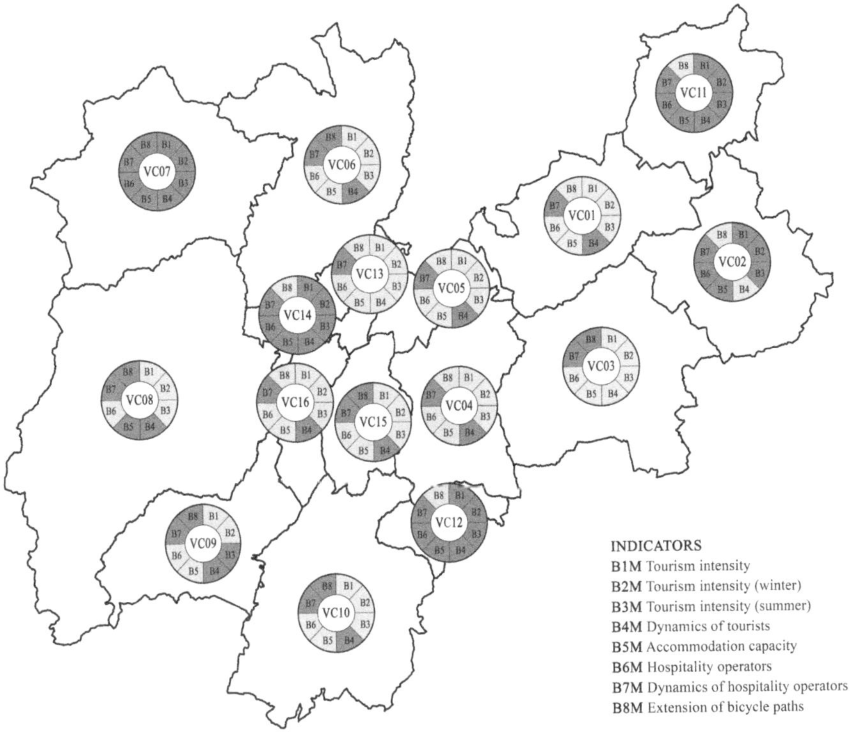

Map 3.4 Tourism: Opportunity and Threat factors in the VCs
Note: In dark grey the opportunity factors, in light grey the threat factors.

by relatively low values in terms of its retail consistency (A1m), specialization (A4m) and typology (A3m). Finally, the cities of Pergine Valsugana and Arco do not present high retail consistency (A1m) although there is a good degree of retail specialization, emerging in Pergine Valsugana, and a varied retail typology (A3m) emerging in Arco.

3.1.2 Tourism: A Region Travelling at Two Speeds

On the macro scale, the analysis and evaluation of the tourist system (theme B – 'Tourism') clearly brings out that there are two distinct situations. On the one hand, the mountain territories are well-established resorts with intense tourism (B1M), mostly ski-related, and marked by a high level of accommodation capacity (B5M) – that is, Primiero, Val di Sole, Val di Fascia, Altipiani Cimbri and Paganella VCs. Tourism is a clear factor for opportunity in these areas. On the other hand, the valley areas and some of the less attractive mountain regions – for example Giudicarie VC – demonstrate a decisively weaker tourist system both in terms of tourism intensity (B1M) of and in terms of accommodation

Map 3.5 Tourism: Strength and Weakness factors in the cities
Note: In dark grey the strength factors, in light grey the weakness factors.

capacity (B5M). Nevertheless, the less tourist-oriented territories are the ones that demonstrate the liveliest dynamics (B4M). This points to the tendency towards a growth in the attractiveness of the more marginal territories. These were the areas that promoted the tourism of eno-gastronomy and excursions as alternatives to the traditional forms of tourism, especially skiing. Several peculiar situations arise inside such a framework that is polarized between highly visited VCs and VCs scarcely touched by flows. Namely, although the Alto Garda-Ledro VC does not demonstrate significant flows of tourists all year long, it enjoys a high-value tourism interest in the summer in relation to the use for swimming and sports activities of Lake Garda and Lake Ledro.

On the micro scale, the tourist system of the five cities that are not in the group of mountain VCs with a strong tourism tradition (quoted above) demonstrates weaknesses in various fields. Above all, Pergine Valsugana and Arco demonstrate this weakness. Here, in fact, the numbers of arrivals (B1m) and nights spent[5]

5 An 'arrival' is defined as every time a guest takes up lodging in the host facility. The guest who makes various stopovers in various host facilities causes more arrivals.

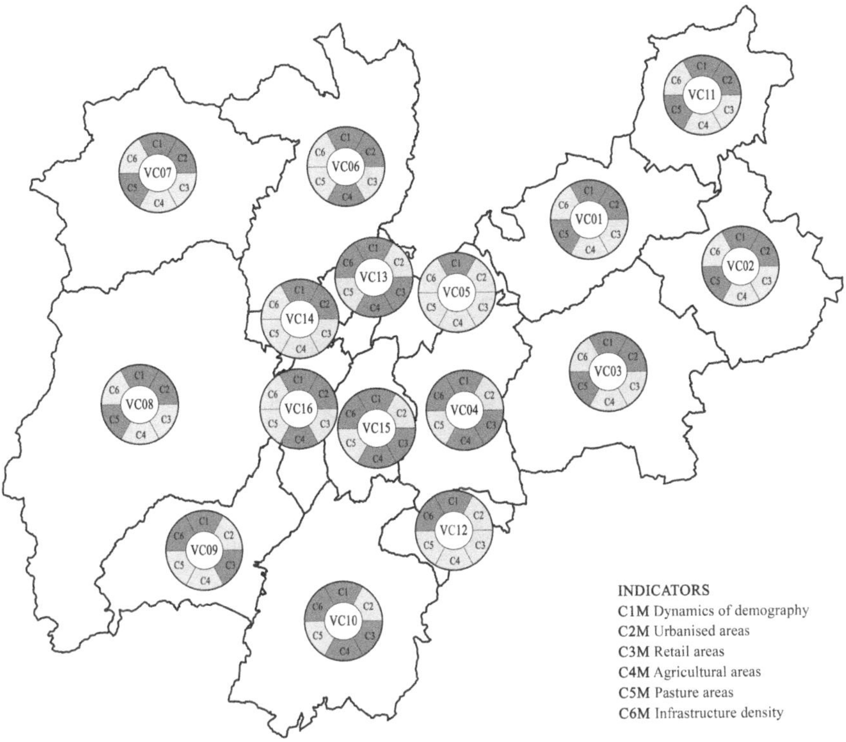

Map 3.6 Land use: Opportunity and Threat factors in the VCs
Note: In dark grey the opportunity factors, in light grey the threat factors.

(B2m) are characterized by below-average values. Instead, Rovereto is marked by a high number of arrivals but by a low number of nights spent, typically of short-time tourism, inspired especially by the cultural offerings, as museums and events. Only Trento and Riva del Garda demonstrate good-level values both in terms of arrival and nights spent, a sign of a tourist system that is a strong point in the local situation, especially at Riva del Garda, a well-known attraction because of its lakeside location.

3.1.3 Land Use: The Urbanized Corridor along the Asta dell'Adige (the Adige River Basin)

The macro scale analysis of the land use (Theme C – 'Land use') revealed that there were two types of VCs. On the one hand, the VCs located along and near the Adige River are marked by a high level of urbanization (C2M), which is

The 'nights spent' are defined as the number of consecutive nights spent by a guest in the same host facility.

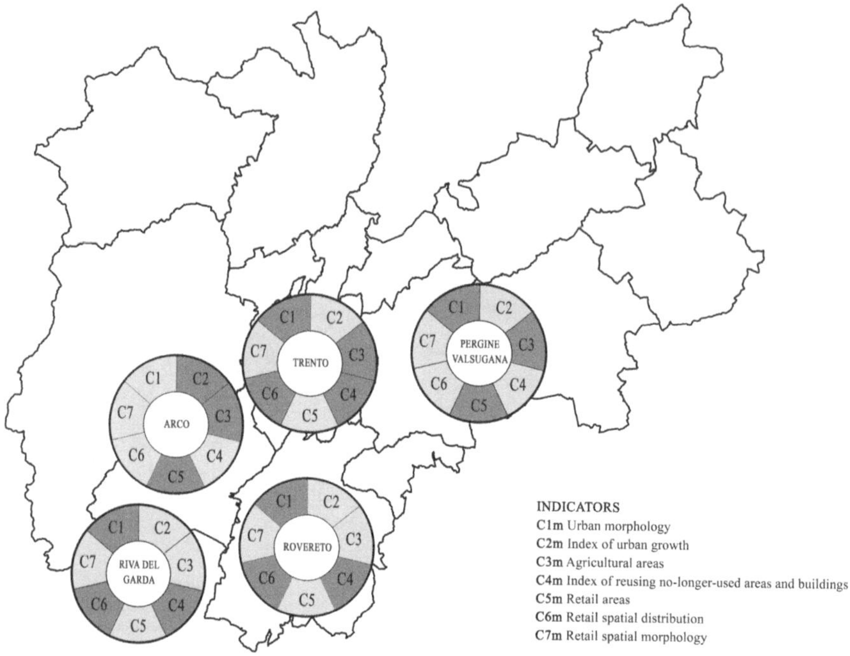

Map 3.7 Land use: Strength and Weakness factors in the cities
Note: In dark grey the strength factors, in light grey the weakness factors.

supported by strongly positive demographic dynamics (C1M): Val d'Adige, Rotaliana-Könisberg, Vallagarina, Alto Garda-Ledro and Alta Valsugana-Bersntol VCs. On the other hand, the remaining VCs are marked by a low level of urbanization and less lively demographic dynamics than the other VCs as well as by a higher incidence of agricultural areas (C4M) and pasture areas (C5M).

On the micro scale, there is a high concentration of disused areas (C4m) that emerge in the cities of Trento and Rovereto. Although these areas are available, the analysis revealed that there is a tendency to extend the urbanized areas (C2m) and hence to seal the soil, especially in the city of Trento. This is a critical phenomenon, if we consider the high incidence of agricultural areas that are subject to the risk of so-called 'artificialization' in the territories of Trento.

3.1.4 Landscape: A Diversified Framework of Values and Risks between Mountains and Plains

The territories of Trentino are endowed with a high variety of geomorphological areas alternating between mountain and valley areas. This is reflected in its history and culture as well as in the variety of local landscapes. The macro scale analysis conducted on the landscape aspects (Theme D – 'Landscape') of the

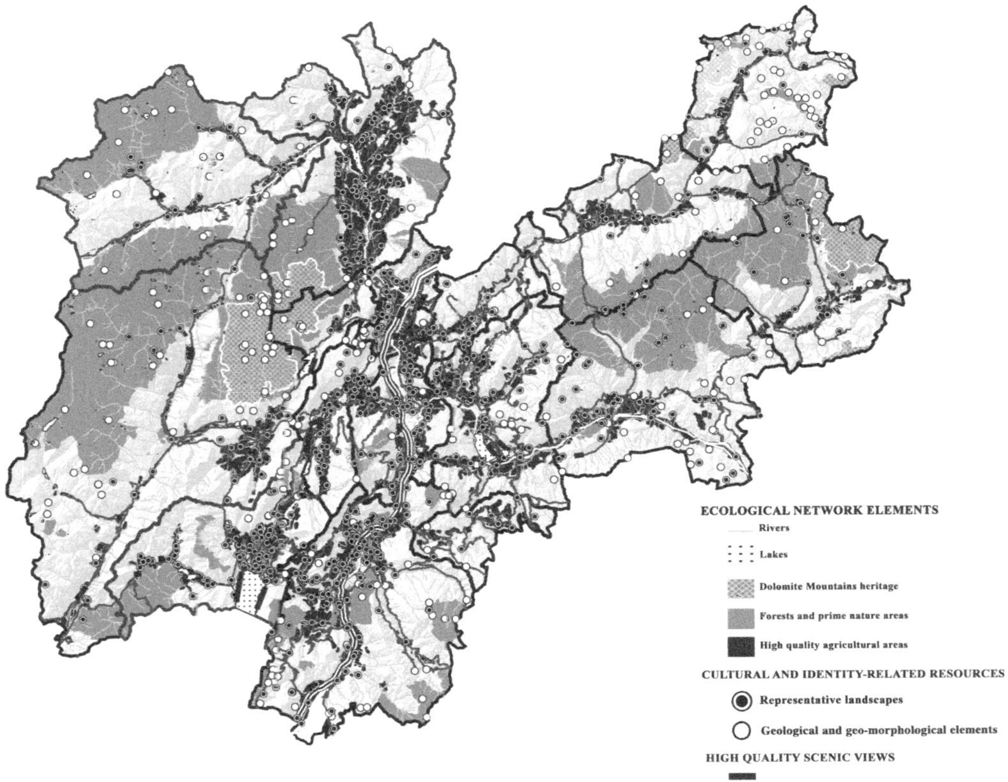

Map 3.8 **Landscape values in Trentino**
Source: Authors' elaboration based on the PTP data.

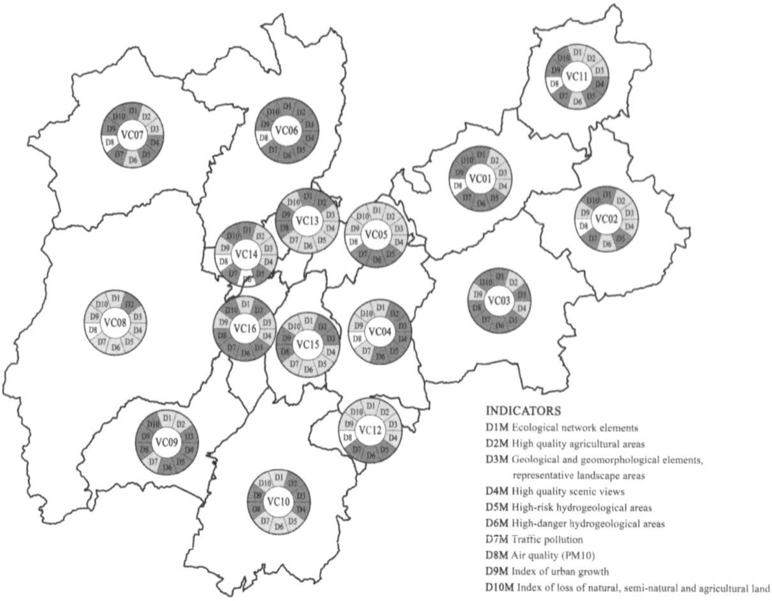

Map 3.9 Landscape: Opportunity and Threat factors in the VCs

Note: In dark grey the opportunity factors, in light grey the threat factors.

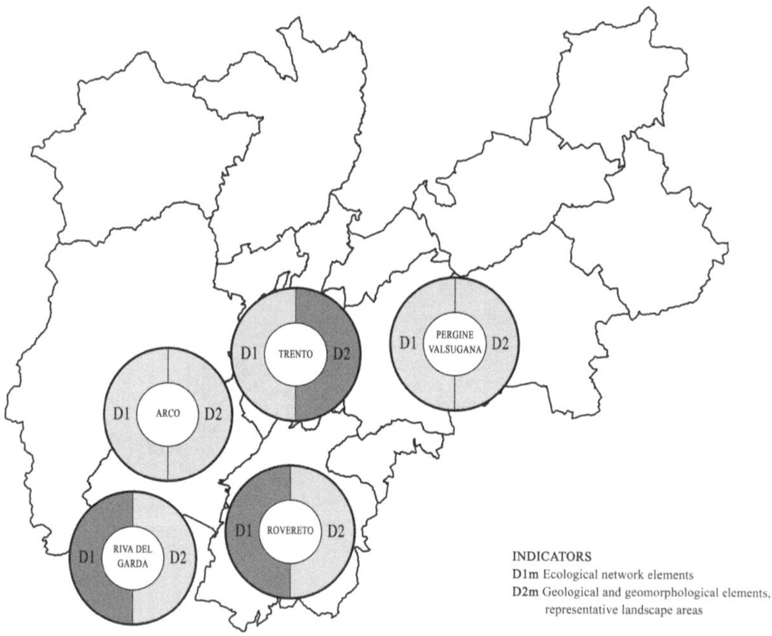

Map 3.10 Landscape: Strength and Weakness factors in the cities

Note: In dark grey the strength factors, in light grey the weakness factors.

Map 3.11 Local governance: Opportunity and Threat factors in the VCs

Note: In dark grey the opportunity factors, in light grey the threat factors, in white indicators non-calculable.

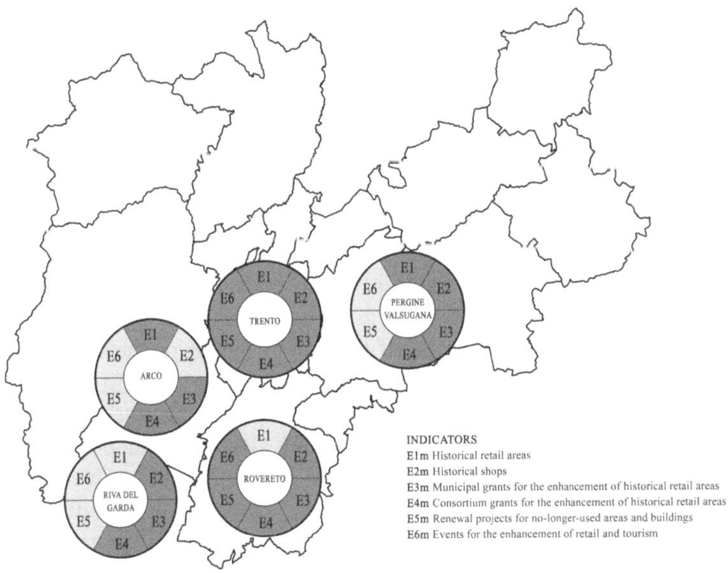

Map 3.12 Local governance: Strength and Weakness factors in the cities

Note: In dark grey the strength factors, in light grey the weakness factors.

mountain areas (Primiero, Val di Fiemme, Val di Non, Giudicarie and Paganella VCs) revealed that these are marked by high naturalistic values, in relation to the presence of protected areas, the resources of the Dolomites, forests, and glaciers (D1M). Instead, the valley areas are marked by landscape values more correlated to anthropic activities, especially agriculture (see the high quality agricultural areas, D2M, in Rotaliana-Könisberg, Val d'Adige, Vallagarina, Val di Non and Alta Valsugana-Bersntol VCs). In addition, the valley areas demonstrate high historical and cultural values, primarily correlated to local resources, especially artistic and architectural resources in the most urbanized areas (D3M). Therefore, the landscape heritage – in its local varieties – consists in an important factor of opportunity for the territories in Trentino.

Despite this opportunity, the landscape heritage is being endangered by energetic processes of urban growth (D9M). This is something that is happening not only in the traditionally urbanized territories – like the Adige River basin – but also in less anthropic environments where the impetus of recent urban development has turned into a dangerous propensity towards the sealing of natural, semi-natural and agricultural soil (D10M).

The micro scale analysis of the landscape in the five cities shows a further articulation of the territorial distribution of landscape values. On the one hand, Trento presents a landscape context marked by relatively low ecological values (D1m) but endowed with a high level of historical and cultural resources (D2m). On the other hand, Riva del Garda and Rovereto are rich in ecological values, but do not have a significant number of historical and cultural resources. Finally, Arco and Pergine Valsugana do not demonstrate specific ecological and historical and cultural values of the landscape within the context of reference of the five cities.

3.1.5 Local Governance: Events as Opportunities for the Revival of Retail and Tourism

The macro scale analysis and evaluation of the level of the initiatives that characterize the local contexts in Trentino (Theme E – 'Local governance') demonstrated that local governance is not directly proportional to the solidity of the retail system or to the liveliness of the tourist system in the territories. This disproportion exists in terms of the enhancement of the traditional retail sites (Provincial grants to cities, E1M, and to consortiums, E2M). In fact, a high level in local governance marks not only the traditionally tourist territories like Altipiani Cimbri and Alto Garda-Ledro VCs but also those that are less attractive from the tourist point of view. This potential constitutes a possible lever for the innovation of both systems – retail and tourism. Furthermore, Val di Fiemme and Val di Fassa VCs have a lively tourist system. Giudicarie, Vallagarina, Rotaliana-Könisberg and Val d'Adige VCs have a good retail system. Nevertheless, these six areas demonstrate a low potential in local governance.

Focusing more closely the theme of local governance on the micro scale, Rovereto and Trento demonstrate that they certainly are the most virtuous cities. These are the cities with a high level of governance in terms of initiatives for the enhancement of traditional retail sites (E3m and E4m), actions taken with no longer used areas and buildings (E5m), and initiatives for the enhancement of retail and tourism (E6m). In contrast, Arco, Riva del Garda and Pergine Valsugana report lower levels of local governance. In particular, there are very few projects for the renewal of no longer used areas and buildings (E5m) or for events for the enhancement of retail and tourism (E6m).

3.2 Scenarios for Local Development

On the basis of the results that came out of the application of the TIE matrix, territorial retail scenarios were defined for each VC. The performances of the individual themes of the TIE in the various territorial contexts were considered in the determination of these scenarios. Two strategic territorial aspects, defined as 'conditions', were also considered, namely, the level of accessibility of the VCs (inter- and intra-Provincial accessibility, in relation to the need for GLAs to be supported by efficient and interconnected transport infrastructures)[6] and the level of integration between retail and other activities ('functional integration', in relation to the opportunity to integrate new formats of retail supply with other urban functions).[7] These aspects are indispensable requirements for the settlement of GLAs in Trentino. In order to check on these two conditions, an overall indicator was developed called the 'index of attractiveness' (see Chapter 5),[8] which enabled us to list a ranking of the VCs that could indicate which territories were most apt for hosting GLAs.

---

6 A good level of car accessibility is a prime consideration in the location of GLAs (Ibrahim, 2002; Kivell and Shaw, 1980).

7 According to Kärrholm (2012), retail should relate to a whole set of other activities (for example tourism, entertainment, culture and leisure). On this, see also Guy (2007).

8 The innovative feature of this indicator lies in the fact that it treats the infrastructures (inter- and intra-Provincial accessibility) and the economy (the correlation of retail and tourism) in an integrated way. In fact, there are very solid analyses of the economic aspects of retail settlement and the growth of GLAs. These analyses consider the importance of the areas of influence of single retail settlements by identifying areas of accessibility bordered by isochrones (see Ghosh and McLafferty, 1987; Gruen and Smith, 1960). This analytical approach is very useful for those who are managing retail distribution for detecting the best location for single retail shops. Nevertheless, it is only a partial technique that cannot alone furnish elements to help evaluate the overall potential of a territorial context to host a GLA. Our aim was to go beyond an analytical approach to this sector, one focused only on the systems of mobility (the flows in and out of the facility, parking, etc.), seeing the GLAs as integrated components of the territorial attractiveness and development. To this end, we worked out the index of territorial

Table 3.1 An exemplifying schema of the strategies, aims and actions identified for one of the 16 VC (Val d'Adige VC)

Scenario	Current Status	Strategy	Aims	Actions
Dominant scenario 1. RETAIL	High index of attractiveness High retail consistency (A1M) Varied retail typology (A7M) and good retail specialization (A8M) Low retail density (A5M)	INTEGRATINGTHE RETAIL SYSTEM	Aim 1 Integrating the existing supply	Action A Creation of new retail structures, even with GLAs, in the city of Trento, through the reuse of no longer used urban areas. Action B Strengthening of the retail supply in the cities of Aldeno, Cimone and Garniga Terme (neighbourhood stores and/ or MRSs)
Prospective scenario 2. MARKETING	RETAIL SYSTEM: SEE THE DOMINANT SCENARIO			
	Positive dynamics of tourists (B4M) Positive dynamic of hospitality operators (B7M) Low level of tourism intensity (B1M) Low accommodation capacity (B5M)	STRENGTHENING OFTHE TOURIST SYSTEM	Aim 1 Construction of a local system of resources Aim 2 Construction of an attractive system Trento-Rovereto Aim 3 Integration between the tourist offerings and the local retail	Action A Networking of existing attractions Action B Increase in the accommodation capacity Action C Promotion of a complementary supply Trento-Rovereto Action D Strengthening of the bicycle paths along the Adige River basin Action E Organization of cultural events, fairs, markets in the city centres

Source: Authors' elaboration.

The integrated interpretation of the results that derived from this ranking and from the results that emerged from the application of the TIE matrix led to the qualification of the various VCs in respect to three main types of scenarios. These are types of scenarios where retail is called to play a different role in relation to the specific potentialities of development of the territorial context. In particular, Scenario 1, 'Retail – Designing the territorial retail system', foresees the settlement of GLAs in territories that are already distinguished by a solid retail system and equipped with the requirements of accessibility and functional integration (alluded to above). Scenario 2, 'Marketing – Designing the integration of retail/tourism/territory', proposes retail as an endeavour that can be integrated profitably with tourism in territories where an especially lively tourist system is in place. Finally, Scenario 3, 'Landscape – Enhancing the landscape identity of the territorial system', envisages the implementation of a type of retail activity that aims to enhance the characteristics of the landscape, closely integrated with the tourist system. Landscape itself is considered as an exceptional 'lever' for the development of retail (see Chapter 2.1.2). This is a scenario that pertains to those territories that already are characterized by high landscape values and a good tourist system.

The three types of scenarios were articulated according to the specific emergent circumstances for each VC. They were identified as a scenario that is dominant, to be interpreted as a leading framework of territorial development in each VC, and/or as a scenario that is prospective, to be interpreted as a potential and future horizon of development. To be implemented, both dominant and prospective scenarios need investments and strategic initiatives promoted by Provincial programming and local planning. Therefore, whether each scenario can be implemented or not is something that depends on the programming choices that Trento Province and the VCs decide to make in agreement with the cities. In order to orient such a process of the implementation of the scenarios in the individual territories, we identified strategies, aims and actions for each VC that were meant to activate the scenario that had been identified (see Chapter 6).

Afterwards, we reported, as examples, a synthesis of the strategies, aims and actions outlined for three VCs. These are VCs that are characterized by different scenarios and that, for this reason, exemplify the variety of characteristics and prospective actions that apply to the territories of Trentino. First, Val d'Adige VC (see Table 3.1) is marked by the dominant Scenario 1, 'Retail – Designing the territorial retail system', as well as by the prospective Scenario 2, 'Marketing – Designing the integration of retail/tourism/territory'. Second, Valsugana-Tesino VC is marked by the prospective Scenario 2, 'Marketing – Designing the integration of retail/tourism/territory'. Third and last, Val di Sole VC is marked by

attractiveness referred to in the text, a complex indicator that can effectively describe a level of attractiveness that is founded on aspects having to do not only with mobility but also with the local productive and economic system.

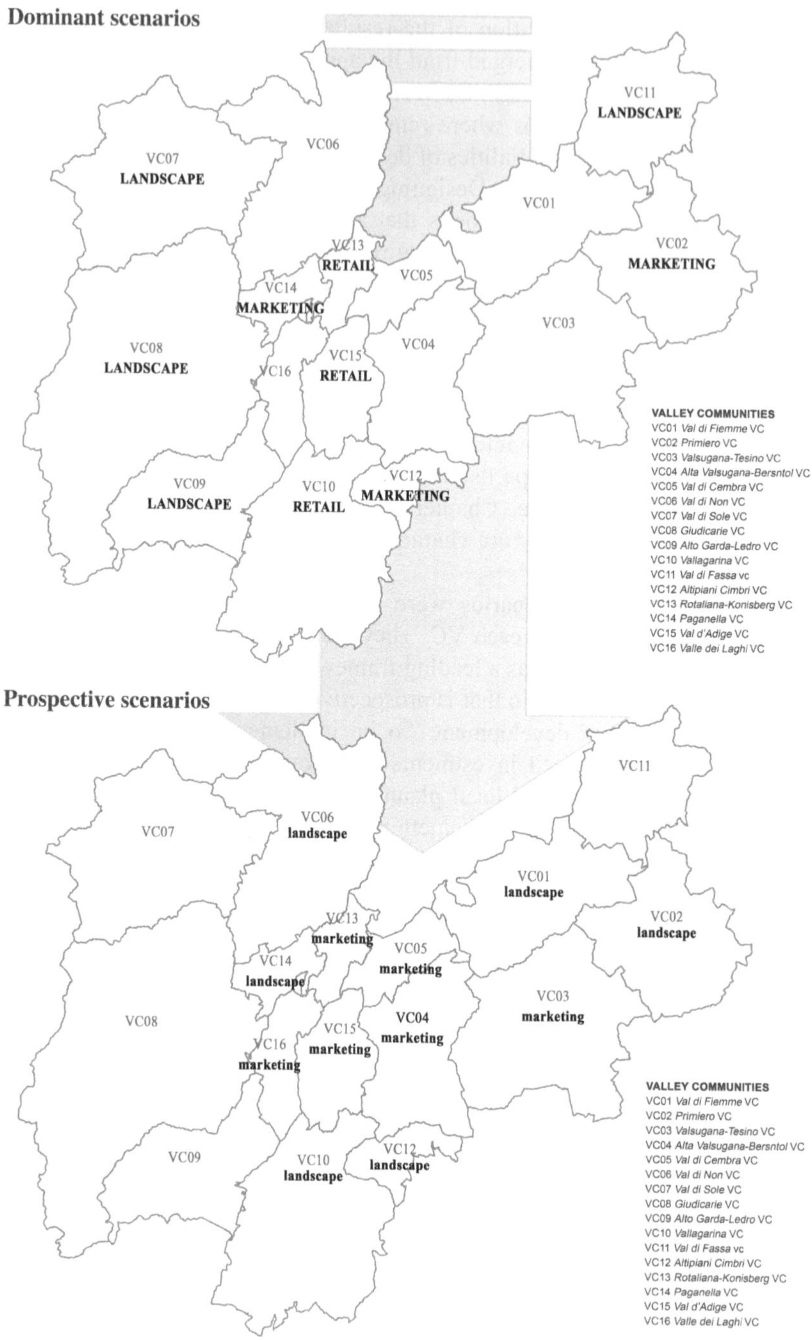

Map 3.13 Territorial retail scenarios in the VCs

the dominant Scenario 3, 'Landscape – Enhancement of the landscape identity of the territorial system'.

3.2.1 Scenarios for Local Development in the Val d'Adige VC

Dominant scenario: Scenario 1, 'Retail – Designing the territorial retail system'
The territorial system of the Val d'Adige VC demonstrates a high index of attractiveness because it is endowed with excellent conditions of inter- and intra-Provincial accessibility (motorway and railway) and with good conditions of functional integration. Its retail system is marked by high consistency (A1M) and low density (A5M). Qualitatively, the territorial system presents a varied retail typology (A7M) and a good retail specialization (A8M).

The dominant scenario for the Val d'Adige VC, 'Retail', needs to be activated by providing for a strategy that aims to integrate the existing retail supply while maintaining today's level of quality in terms of varied retail typology and specialization. In the context of Val d'Adige VC, there are essentially two complementary ways for activating this scenario. First, there should be the creation of new retail structures, even GLAs, in the city of Trento, through the reuse of disused urban areas. In particular, the new retail settlements must address themselves to new typology appropriate for the integration of various functions, for example, entertainment and free time. Moreover, they must be characterized by a retail supply diversified and complementary to the supply already existing. Second, there should be a strengthening of the retail supply in the cities of Aldeno, Cimone and Garniga Terme (neighbourhood stores and/or MRSs), which have evidenced decreasing retail dynamics (A4M).

Prospective scenario: Scenario 2, 'Marketing – Designing the integration of retail/tourism/territory'
In the context of Val d'Adige VC, the prospective scenario, 'Marketing', aims to integrate retail supply – after it has been completed and integrated according to the above cited strategies of the dominant scenario – with local attractions. In this perspective, the positive dynamics in recent years (2001–2010) of tourists (B1M) and of accommodation capacity (B5M) constitute an important factor of opportunity for the Val d'Adige VC. This VC's tourist system suffers from the competition of the more touristic VCs in Trentino– Alto Garda e Ledro, Altipiani Cimbri, Val di Fiemme and Val di Fassa VCs to the east; Giudicarie, Paganella and Val di Sole VCs to the west. Nevertheless, it has demonstrated that its tourism is growing. Therefore, if strengthened, tourism could give a precious lift to the definition of a scenario of tourism–retail integration.

In order to activate the prospective scenario, the VC needs to provide for a strategy for the construction of a local system of resources. The existing attractions should be networked, above all, the artistic and historical resources, the museums and the cultural events. This operation must then be accompanied by a substantial increase in the accommodation capacity, which is now one of the lowest in the Province.

This system can be put together much more effectively if it is placed within a broader strategy of the enhancement of tourism that takes in the nearby Vallagarina VC. In fact, the aim must be that of putting together a tourist system that is not just local but territorial, one that has two complementary and strictly interconnected fulcrums in the two main urban centres, Trento (Val d'Adige VC) and Rovereto (Vallagarina VC). The complementary nature of this system lies inside the system of cultural resources (with special reference to the Science Museum – MuSe in Trento and the Museo di Arte Moderna e Contemporanea in Rovereto, conceived as the main centres of attraction). In addition to this, the system can be complementary in a more general way in that it could provide a tourist system in Val d'Adige VC that is more focused on cultural resources and a tourist system in Vallagarina that is more oriented towards an integration of cultural and natural resources. What can be foreseen is the construction of a 'city of Adige', founded on the two centres – Trento and Rovereto – and equipped to supply integrated tourist offerings. In this prospective, the network of bicycle paths, that currently runs along the Adige River connecting Trento and Rovereto, could play a role of strategic importance.

Finally, the prospective scenario envisages the integration between the renewed tourist offerings of the VC with its local retail. For example, events occurring in the centres of cities, especially Trento, can be promoted (such as fairs and markets).

3.2.2 Scenarios for Local Development in the Valsugana-Tesino VC

Prospective scenario: Scenario 2, 'Marketing – Designing the integration of retail/tourism/territory'

The territorial system of the Valsugana-Tesino VC demonstrates a low index of attractiveness. This system is marked by low level of accessibility (inter- and intra-Provincial accessibility) and low level of functional integration. As far as the VCs' retail system is concerned, it is characterized by low consistency (A1M) and density (A5M), but it demonstrates a good variety of retail typology (A7M) and specialization (A8M).

In order to implement the prospective scenario, 'Marketing', a strategy of integration of the VCs' retail supply must be followed. Such a strategy must proceed through the opening of neighbourhood stores and/or MRSs, especially in the areas characterized by low retail consistency, that are the scarcely populated areas and areas where there is a risk of retail desertification. In addition, forms of retail management, such as the already existing multi-service shops[9] or the door-to-door selling and the ethical purchasing groups, should be strengthened and promoted in order to mount a defence of retail in the VC. In the other areas, specialized retail supply, such as the sales of local agricultural and dairy products as well as craft articles, should be promoted. These retail initiatives should give preference to the settlement of shops inside or near the city centres.

Furthermore, in order to implement the strategy of integration and improvement of the quality of the VCs' retail supply, we need to keep in mind the retail shops foreseen today in the VC that have already been authorized but not yet built. These include several MRSs to be located in Borgo Valsugana as well as two GLAs foreseen to be located in Novaledo and Borgo Valsugana. In relation to these GLAs, they should promote offerings that are different from those already existing in the city centres of Borgo Valsugana and Novaledo in order to avoid competition between the new large structures and existing shops (Teller, 2008). Synergetic actions could be developed and shared by the shops in the city centres and the GLAs: both non-material actions, such as the promotion of coordinated events, and material actions, such as the sharing of parking lots.

While the retail supply is being increased, the already existing retail supply should also be improved, especially that situated in city centres. To do this, the cities could take initiatives to enhance the historical retail areas.[10] In addition, the cities in the urban areas could organize events for the integrated improvement of retail and tourism, correlated, for example, with the promotion of local products. To this end, it would be advisable that retail operators, agricultural personnel and craftspeople get together in forms of joint management for promoting and marketing local products.

To implement the 'Marketing' scenario, a strategy should be devised and also followed to strengthen the local tourist system. At present, the VCs' tourist system is characterized by a rich panorama of highly diversified resources: the Massif of the Lagorai Mountains, the Brocon ski areas, the Tesino golf courses, the baths of Roncegno, the bicycle paths, the high quality agricultural areas, as well as cultural resources. Nevertheless, this panorama of attractions comes across as very fragmented and little attractive overall, especially if it is contrasted with other tourist territorial systems in Trentino. Thus tourism for this VC is

9 The multi-services shops are defined as those stores that, beyond the principal activity of providing basic goods, also provide other services, such as a postal service, a cash point, a tourist information point and an internet connection (Provincial Law 17/2010, exhibit b).

10 In recent years, retail turns out to be an element that can support the project of urban regeneration, not only from a purely economic point of view but also from a social and territorial standpoint, by providing jobs, services, investments and a focal point for community activities. In this perspective, the literature offers a wide range of examples on the crucial role of retail in successful revitalization processes of city centres (Balsas, 2001; Brunetta and Caldarice, 2014; Gibbs, 2012; Mitchell and Kirkup, 2003).

a potential that still needs to be expressed and managed through systematic policies that would promote the various existing resources in an integrated way. Hence, the 'product Bassa Valsugana VC' should be based on the networking of the diversified resources cited above for the promotion of a kind of tourism that is multifaceted: excursion and sports tourism, rural and gastronomic tourism, and cultural tourism. Besides this, the VC should increase its accommodation capacity, taking measures to favour *agriturismo* facilities (local farm-food and lodging facilities) and developing projects for the renovation of traditional rural farmhouses (the *maso*) and mountain chalets (the *baita*). It is also important to complete the network of bicycle paths that currently runs along only few valley areas, in order to create an effective link with the nearby Alta Valsugana-Bersntol VC. This link could foster the development of a unique tourist district made up by Valsugana-Tesino VC and Alta Valsugana-Bersntol VC. The so-renewed local tourist system could be effectively integrated into the above cited specialized retail supplies.

3.2.3 Scenarios for Local Development in the Val di Sole VC

Dominant scenario: Scenario 3, 'Landscape – Enhancement of the landscape identity of the territorial system'

The territorial system of the Val di Sole VC demonstrates a low index of attractiveness, in that there is a low level of inter- and intra-Provincial accessibility and there is very little functional integration. The system of local retail supply is marked by low retail consistency (A1M) but good density (A5M). Qualitatively, retail does not perform well in terms of typology (A7M) and specialization (A8M). In order to activate the dominant scenario, 'Landscape', there should be a retail strategy that has two aims. First, the strategy should try to guarantee the permanence of retail shops in the cities that risk retail desertification. Second, the strategy should try to integrate the retail system and improve its quality in terms of typology and specialization, providing for new retail shops, preferably neighbourhood stores and MRSs. These new retail shops should prefer types of specialized offerings that are correlated with local products – agricultural dairy and crafts – and with the sports activities practiced and events organized in the VC. They should be situated inside or near urban areas in view of enhancing the historical city centres. The promotion of a specialized supply of products correlated to local ones can be also made through the distribution of characteristic products in GLAs thanks to special forms of retail promotion (such as corners or stands) and through the organization of temporary forms of retail (such as farmers' markets).

The main strategy in relation to the local tourist system consists in the maintenance of its already high quality, also to guarantee its current high number of winter and summertime tourists (B2M, B3M) in the future and keeping up its already good level of accommodation capacity (B5M). However, this objective of maintaining the characteristics of success of the local tourist system should be sought through diversifying the tourist offerings more and more, emphasizing the value of the multifaceted resources that the VC can count on: skiing facilities, which have a very high number of users, especially at Folgarida and Marilleva; thermal bath centres at Pejo and Rabbi; protected natural areas, including the Adamello Brenta Provincial Nature Park and the Stelvio National Park; and the various areas of the VC where people can engage in sports activities. To this end, the bicycling circuit has become important. It needs to be completed along the side valleys of the Val di Sole VC. It could be integrated effectively with the Trento-Malè railroad line, so as to create a local system of integrated train–bicycle mobility that could also be connected to the nearby Val di Non VC.

The objectives of the integration of tourist and retail offerings can be pursued through the promotion of local products, such as the apple (DOP, that is, an origin guaranteed food), which is grown in the milder climatic areas of the VC and 'shared' with the neighbouring Val di Non VC, and the promotion of activities like excursions, river sports, winter sports and cycling. In fact, the Val di Sole VC has already been the site of internationally famous sports events. This promotion could be correlated with highly specialized retail shops.

There is finally need for the activation of a strategy for the conservation of the present landscape values, which are exceptionally high, above all, ecologically (D1M) and scenically (D2M). This should be done through monitoring the potential spread of urbanized areas and therefore of soil sealing, especially in relation to the risk of the proliferation of vacation homes. This is a very critical phenomenon, especially when we consider the high level of hydrogeological danger that characterizes the territorial context of the VC (D6M) – the potential avalanches and landslides. In reaction to this resident-tourist pressure, a priority strategy should be the renovation of what already exists, including not only the historical and traditional buildings but also the large facilities for hospitality. The protected natural areas inside the VC – a Provincial (Adamello Brenta) and a National Park (Stelvio) – constitute important opportunities for conserving and enhancing the context of the local landscape, not only inside the parks' protected boundaries but also outside, developing synergy between administrations inside and outside the Parks.[11]

11 According to the so-called 'new paradigms' of nature conservation (Crofts, 2009; Phillips, 2003a; 2003b), ratified on the occasion of the Vth IUCN World Park Congress held in Durban in 2003, a prime mission of Protected Areas (PAs) is that of extending the benefits of their policies beyond their institutional boundaries, towards their territorial context. This extension is beneficial for both the PAs, which are in this way preserved by the critical phenomena that usually originate in their territorial context, and for the surrounding areas, which can share the positive effects of PAs nature conservation policies (Salizzoni, 2012).

References

Balsas, C., 2001. Commerce and the European city center: Modernization, regeneration and management. *European Planning Studies*, 9(5): 677–82.

Brunetta, G. and Caldarice, O., 2014. Self-organisation and retail-led regeneration: A new territorial governance within the Italian context. *Local Economy*, 29(4–5): 334–44.

Crofts, R., 2009. Protected areas, from Durban onwards. *IUCN Parks Magazine*, 17(2): 5–12.

Ghosh, A. and McLafferty, S.L., 1987. *Location Strategies for Retail Firms and Service Firms*. Lexington, MA: Heath.

Gibbs, R.J., 2012. *Principles of Urban Retail Planning and Development*. Hoboken, NJ: Wiley.

Gruen, V. and Smith, L., 1960. *Shopping Town USA: The Planning of Shopping Centers*. New York: Reinhold.

Guy, C.M., 2007. *Planning for Retail Development*. New York: Routledge.

Ibrahim, M.F., 2002. Disaggregating the travel components in shopping center choice: An agenda for valuation practices. *Journal of Property Investment and Finance*, 20(3): 277–94.

Kärrholm, M., 2012. *Retailising Space: Architecture, Retail and the Territorialisation of Public Space*. Farnham: Ashgate.

Kivell, P.T. and Shaw, G., 1980. The Study of Retail Location. In: J.A. Dawson (ed.), *Retail Geography*. New York: Wiley, pp. 95–155.

Mitchell, A. and Kirkup, M., 2003. Retail development and urban regeneration: A case study of Castle Vale. *International Journal of Retail and Distribution Management*, 31(9): 451–8.

Phillips, A., 2003a. A modern paradigm. *World Conservation*, 2: 6–7.

Phillips, A., 2003b. Turning ideas on their head: The new paradigm of protected areas. *The George Wright Forum*, 20(2): 8–32.

Salizzoni, E., 2012. *Paesaggi protetti. Laboratori di sperimentazione per il paesaggio costiero euro-mediterraneo*. Firenze: FUP.

Teller, C., 2008. Shopping streets versus shopping malls: Determinants of agglomeration format attractiveness from consumers' point of view. *The International Review of Retail, Distribution and Consumer Research*, 18(4): 381–403.

Chapter 4
The Analytic Network Process Method for Territorial Integrated Evaluation

Patrizia Lombardi

4.1 Multi-Criteria Analysis in Spatial Planning

Decision-making at different spatial planning levels is usually concerned with the evaluation of alternatives and the selection of an action to take. The alternatives can also vary in type and intensity of their effects. This process can be improved through the use of Multi-Criteria Analysis (MCA), which provides a systematic process for trading off the effects of various alternatives, taking into account all the aspects and values involved in decisions and synthesizing individual contributions (Saaty, 2000; Saaty and Ozdemir, 2008; Saaty and Vargas, 1994).

Originally developed in the field of Operative Research, discrete MCA is able to draw up lists of priorities from a finite series of choice options (or alternatives) that are assessed and compared in relation to the identified characteristics of the problem (or criteria) when a characteristic is appropriately broken down into its fundamental elements (Voogd, 1983). In particular, MCA provides the following benefits:

- a clear definition of the criteria used in the selection of an option between alternative solutions; a weighting of the criteria to be used in the evaluation, in accordance with different points of view;
- a combination of multiple aspects which have different natures; a comparison among the objectives and strategies of the various subjects involved and the available resources;
- a transparent and explicit evaluation approach. By taking into account different points of view and a weighting of the criteria to be used in the evaluation of alternative options, MCA gives decision-makers a chance to make a meaningful and pluralistic evaluation of planning proposals, synthesizing all the contributions of the different experts and the points of view of the actors involved (stakeholders and decision-makers).

In the field of planning and evaluation, most of the MCA methods that are applied are 'discrete' methods since these methods are able to manage a limited number of alternatives, corresponding to few projects as well as multiple indicators, quantitative and qualitative in nature. Decision-makers can express their views, assigning individual preferences to the various criteria of evaluation. Therefore, discussions and negotiations should be encouraged where various decision-makers may be represented – exponents of different groups of opinion, political currents and lobbies, as well as the promoters and executors of the proposed actions (Brandon and Lombardi, 2011; Lombardi, 2009; Miller and Patassini, 2005).

The literature presents a wide range of MCA methods which can be grouped in families – quantitative, qualitative and mixed (Deakin and Lombardi, 2005; Greco et al., 2001; Lombardi, 2007; Nijkamp, Rietveld and Voogd, 1990; Roy and Bouyssou, 1995; Saaty, 1995; Voogd, 1983). These differ from each other in the nature of information they are able to manage – that is, cardinal (hard), ordinal (soft) or mixed data.

MCA methods are subdivided further according to the systems they use to treat preferences. There are 'compensatory methods', such as the methods of multi-attribute utility – Multiple Attribute Utility Theory (MAUT), Analytic Hierarchy Process (AHP) and Analytic Network Process (ANP); 'partially compensatory methods', such as the methods of REGIME, PROMETHEE, and NAIADE; and 'non-compensatory method's, such as the lexicographic method and the method of dominance (Bottero and Ferretti, 2011; Bottero, Lami and Lombardi, 2008; Figueira, Greco and Ehrgott, 2005; Munda, 2005).

Inside the large 'family' of MCA (Figueira, Greco and Ehrgott, 2005), the ANP is the only decision-support method which makes it possible to deal systematically with all kinds of dependencies and feedback. ANP is an advanced version of the AHP, developed by Saaty (1980; 1992; 1993; 1994; 1995; 1996a; 2001), which translates expert judgements into a nine-point scale, providing cardinal indexes for operationalizing (Saaty, 2005).

The ANP model consists of clusters (that are groups of homogeneous elements of a decision issue), elements (that are nodes of the network), interrelationships among clusters, and interrelationships among elements. Relations can be bidirectional (that is cause-effect) or mono-directional (that is simple influence without reciprocity).

This method better represents the reality of complex systems like those of the urban environment and its transformations. It allows interactions and feedback within and among clusters, providing a process from which ratio-scale priorities can be derived from elements (Saaty, 2005).

The ANP offers a more accurate and realistic representation for policy-making support because it describes the interaction between elements belonging to heterogeneous clusters in a complex system more appropriately than other methods. For this reason, the ANP has been applied to a variety of decision problems in different fields, including the prediction of the outcomes of competitions in business and sports (Saaty, 1996b).

In this study, the ANP has been applied to Territorial Integrated Evaluation (TIE) for the management of the territorial retail development in Trentino (see

Chapter 3). The TIE methodology was reinforced and augmented using the ANP in order to model, cluster and begin evaluating possible spatial scenarios according to a 'what-if framework'.

In the TIE methodology, the ANP model is developed at both the macro (Valley Community) and the micro (city) scale, comparing the indicators' priorities to the relative control criterion for a better definition of territorial scenarios (see Chapter 1). In particular, the ANP method enables decision-makers to select the most relevant indicators for defining the spatial scenarios, which are obtained through an appropriate synthesis formula.

4.2 The Analytic Network Process Methodological Approach

The ANP (Saaty, 1996b; 2005) is a generalization of the AHP. The mathematician Thomas L. Saaty initially developed the AHP, and later the ANP model to help managers make decisions (Saaty, 1980; 1995). Subjective assessments and objective facts are incorporated into a logical-hierarchical AHP framework or into a network ANP framework to provide decision-makers with an intuitive and common-sense approach in quantifying the importance of each decision element through a process of comparison. This process enables decision-makers to represent a complex problem in a hierarchical or in a network form (Gomez-Navarro et al., 2009; Lombardi, 2010; Saaty and Forman, 1993). It allows the investigation to reach a set of ratings for the decision alternatives by aggregating the relative weights of decision elements.

Nevertheless, AHP uses a linear hierarchical structure. In it, the relations of dependence among the elements of the various decisional levels are unidirectional along the hierarchy. Furthermore, there are not any relations of dependence among the elements (also-called 'nodes') of the same homogeneous group of elements (also-called 'clusters'). There are no relations of dependence among elements from different clusters. In contrast, ANP uses a network to represent the complexity of the real decisional issue, a network that presents relationships of internal and external dependence among the elements of the decisional system.

Concisely, the ANP methodology involves the following steps (Saaty, 2001; 2005):

- Structuring the decision-making model. This activity involves an identification of both the elements constituting the decision issue and their relationships. The network model consists in various 'clusters' (that are groups) of elements (that are 'nodes'), and alternatives or options from which to choose. Each element can have influence relations and inter-dependence relations: each element can be a 'source' – that is an origin of a path of influence – or a 'sink' – that is, a destination of paths of influence';

- Developing paired comparisons of both elements and clusters to establish relations within the structure. In this step, a series of paired comparisons are made by participants in the decision-making process (usually experts, managers and representatives of citizens) to establish the relative importance of decision elements with respect to each component of the network. In paired comparisons, a ratio scale of number one to nine is used, named Fundamental Scale or Saaty's Scale (see Table 4.1). The numerical judgements established at each level of the network form pair-matrices which are used to derive weighted priority vectors of elements (Saaty, 2001);
- Achievement of the final priorities. To obtain the global priority vector of the elements, including the alternatives, the mathematical approach encompasses the use of 'supermatrices' (portioned matrices composed of sub-matrices consisting of the priority-weight vectors of the elements which have been evaluated). A final supermatrix is obtained at the end of the process, containing the global priority vector of the elements.

The above steps are better illustrated hereafter.

4.2.1 Structuring an ANP Decision-Making Model

In the AHP, the procedure starts by breaking down the decision problem into a hierarchy of interrelated decision elements. At the top of the hierarchy lies the most macro decision objective, that of selecting the best alternative. The lower levels of the hierarchy contain attributes which contribute to the quality of the decision. Details of these attributes increase at the lower levels of the hierarchy. The last levels of the hierarchy contain decision alternatives.

In the ANP, instead, a network model is used to represent the decision-making problem. As previously stated, the ANP more appropriately describes the interaction among elements belonging to heterogeneous clusters in a complex system. The reason is that many decision problems cannot be structured hierarchically because they involve the interaction and dependence of higher-level elements on lower-level elements. In other words, the feedback structure does not have the linear top-to-bottom form of a hierarchy but looks like a network, with cycles connecting its clusters of elements and with loops that connect a cluster to itself.

Usually in a network model, not only does the importance of the criteria determine the importance of the alternatives but also the importance of the alternatives themselves determines the importance of the criteria. However, if there is no feedback and if decision-makers wish to diminish complexity, the alternatives generally can be excluded and the influence among the remaining clusters may be examined.

There are two possible modelling approaches to ANP. One is the Benefits, Costs, Opportunities, Risks (BOCR) approach, suggested by Saaty (Saaty and

Vargas, 2006; Wijnmalen, 2007), which helps simplify problem-structuring by classifying issues into traditional categories of costs and benefits. The other is a simple network, a free-modelling approach, which may better reflect the complexity of a problem.

The first approach is the so-called BOCR model or Strengths, Weaknesses, Opportunities, Threats (SWOT) model. This model assumes that a hierarchy of control exists, one that generates sub-networks that contain groups, elements and alternatives. This is the model that will be used in the case study illustrated in this chapter.

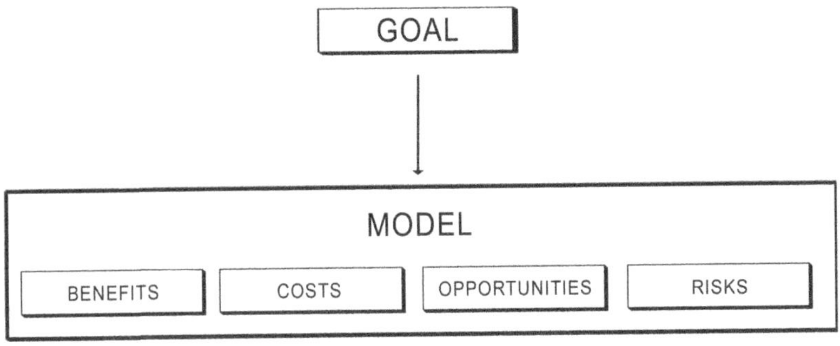

Figure 4.1 The BOCR model
Source: Author's elaboration. All subsequent figures are from the same source, unless otherwise stated.

The multiplicity of factors that characterize the analysis can be simplified as long as every decision issue is seen as characterized by favourable and unfavourable aspects. The favourable aspects can be considered 'benefits' while the unfavourable aspects can be considered 'costs'. The very uncertainty of every decision can take on positive characteristics and hence become an 'opportunity' or this uncertainty can take on negative characteristics and transform itself into a 'risk'.

Each of these factors can generate a sub-network in which the various elements at play can be evaluated, so that they then can be integrated later in order to take their places in a ranking of the final extent of preference of the alternatives.

In addition, the BOCR (or SWOT) model can be called hierarchical because, at its first level, it presents a structure that is typical of AHP. Its set up brings along greater limitations because it cannot have any of the numerous connections typical of a network structure. Despite this, the structured-network model enables decision-makers to handle a complex problem, simplifying the problem and giving rise to a fitting number of issues while facilitating numerous connections among nodes and clusters.

To create a network model one should:

- Specify the aim of the analysis;
- Describe the alternatives – that is, the different answers to the issue to be decided;
- Dismantle the parts of the problem, identifying the various clusters and their relative nodes in every sub-network (benefits, opportunities, costs and risks). The same data will later be laid out in the sub-networks inside the ANP model and used to express judgements that contrast pairs, depending on their relations of dependence among one another (Farkasovsky and Greda, 2006).

Under each criterion or sub-criterion in every control hierarchy, a hierarchy or a network is developed that is appropriate for that criterion. The same set of alternatives must be included in all these structures. The following steps are suggested (Saaty, 1996b):

- Perform four types analyses: one for benefits, one for opportunities, one for costs and one for risks;
- In each, use a control hierarchy of criteria and sub-criteria to separate the kinds of interactions;
- For each of the criteria or sub-criteria, create a hierarchy or network of the different clusters. Identify the elements in each cluster and draw arrows connecting clusters, indicating which cluster affects which other cluster. If there is inner dependence, draw a loop for that cluster.

In the ANP procedure for structuring the issue to be decided, users should rely on the information and understanding available to develop a control hierarchy – benefits, opportunities, costs, and risks – and then combine their results for the alternatives in the usual way.

4.2.2 Developing Pairwise Comparisons

After users have finished constructing the model, they can go on to make value judgements about each decision element that makes up the model, doing this through the comparative method.

This method consists in comparing pairs of elements in every cluster. These pairs, as stated above, turn out to be homogeneous. They can be compared with one another in reference to each parent node – that is, in respect to each element with which there turns out to be any type of dependence, singled out by this criterion.

Taking into consideration the relationship of dependence hypothesized in the construction of the decision model, every element in the network can take its turn as a 'parent' and be used as a term of reference for the paired comparisons between the 'offspring' nodes connected to it.

The evaluation takes place through paired comparisons and is developed on two levels: the level of the clusters and the level of the nodes. Both types of comparison are dependent on the connections established previously among the nodes of the network.

The process of comparison is represented by forming a comparative matrix (see Figure 4.2). Pairwise comparison data are collected for only half of the matrix elements: diagonal elements always equal one, and the lower triangle elements of the matrix are the reciprocal of the upper ones.

m	*1*	*2*	*3*	...	*n*
1	1			Giudizi	
2		1			
3			1		
...		Reciprocal numbers		1	
n					1

Figure 4.2 Paired comparisons matrix

Pairwise comparisons give users a basis for revealing their preferences through comparing two elements. Furthermore, users have the option to express their preferences between the two to these degrees: 'equally' preferred, 'weakly' preferred, 'strongly' preferred or 'absolutely' preferred. These can be translated into paired weights of, 3, 5, 7 and nine, respectively. The numbers 2, 4, 6 and 8 are used as intermediate values when there is not agreement between preferences. The reciprocal numbers 1–2, 1–3 … 1–8, 1–9 complete the matrices (see Table 4.1).

This ANP technique takes the above comparisons as its input and produces the relative weights of elements at each level as output, using the 'eigenvalue' method. The eigenvector of each comparative matrix is the priority list. Meanwhile, the eigenvalue gives the measure of consistency in making the assessment or comparison. The synthesized eigenvector is the global sequence of the alternatives with respect to achieving the goal. The eigenvector and the

Table 4.1 Saaty's fundamental scale

Intensity	Definition	Meaning
1	Equal importance	Two activities/elements contribute equally to the objective
3	Moderately important	Experience and judgement slightly favour one activity over another
5	Strongly Important	Experience and judgement strongly favour one activity over another
7	Very strongly important	An activity is favoured very strongly over another; its dominance demonstrated in practice
9	Extremely important	The evidence favouring one activity over another is of the highest possible order of affirmation
2, 4, 6, 8	Intermediate values	

Source: Author's elaboration based on Saaty (1980).

maximum eigenvalue of the comparative matrix are determined by solving the general problem of eigenvalues (equation 1):

$$AW = \lambda_{max} W \ (1)$$

Where:

A is the comparative matrix
W is (W1, W2, W3, W4, W5)T eigenvector
λ_{max} is the maximum eigenvalue

The ANP has a systematic procedure for better judgements (Saaty, 1996). This is based on the calculation of the Consistency Index (CI) of each matrix of comparisons. This index provides evidence of the consistency of the judgements given by the ANP users. As shown in the equation 2, the CI is given by:

$$CI = \frac{\lambda max - n}{n-1} \ (2)$$

Finally, a Consistency Ratio (CR) is obtained by comparing the CI with the appropriate one of the set of numbers shown in Table 4.2, each of which is an average random consistency index derived from a large sample of randomly generated reciprocal matrices using the scale 1–9, 1–8 … 1 … 8, 9. The resulting vector is accepted if CR is about 0.10 or less. For n=3, it is 0.05 and for n=4, it is 0.08. If it is not less than 0.10, it is necessary to study the problem and revise the judgements.

Table 4.2 The Random Consistency Index (RCI) for different size matrices

Size matrix – n	1	2	3	4	5	6	7	8	9	
Random Consistency Index		0	0.5	0.8	1.1	1.2	1.3	1.4	1.4	1.4
			2	9	1	5	5	0	5	9

Source: Author's elaboration.

4.2.3 Achievement of the Final Priorities

In order to achieve the final priorities, ANP makes use of supermatrices. The numerical values or 'local priorities' that result from the paired comparisons among the various elements of the clusters constitute a non-weighted 'supermatrix'. The supermatrix contains all the influences of the elements that constitute the network in a concise form.

In the first place, the supermatrix contains within itself the various blocks of relations of dependence and inter-dependence, blocks which take the form of autovectors. Therefore the reading of the supermatrix enables users to understand the relations of influence determining the development of the decisional network.

In the second place, the supermatrix, because it is made up of various autovectors, furnishes numerical indicators about the priorities of the elements that are part of the decision system.

The generic form of the supermatrix is described in Figure 4.3. Here, 'CN' represents the Nth cluster. 'Nn' represents the nth element inside the Nth cluster. 'W_{ij}' is a block of the matrix that contains the priority vectors (w) of the influence of the elements of the i-th cluster in respect to the j-th cluster.

Each column of the matrix W_{ij} is constituted by the main autovector of the corresponding matrix of paired comparisons. This autovector represents the influence of the nodes in the i-th cluster of the network on a node in the j-th cluster. If the i-th cluster does not exert any influence on the j-th cluster, then the matrix W_{ij} is constituted only by zero.

As we have seen in the development of ANP, an 'initial or non-weighted' supermatrix was created, composed of the priority vectors (w), obtained by multiple paired comparisons developed over the course of the analyses. In addition, a weighted supermatrix is created. This also takes into account the priorities attributed at the level of the comparison among the different clusters of the analysis. This supermatrix is obtained by multiplying the values of the non-weighted matrix by the weight of every cluster.

As seen, in fact, the first step in the decisional model is constituted by the compilation of the paired-comparison matrices on the level of the clusters. As a result of this paired comparison, therefore, the priority vectors of the clusters are

$$
W = \begin{array}{cc} & \\ C_1 & \begin{array}{c} e_{11} \\ \cdots \\ e_{1n1} \end{array} \\ C_2 & \begin{array}{c} e_{21} \\ \cdots \\ e_{2n2} \end{array} \\ \vdots & \\ C_N & \begin{array}{c} e_{N1} \\ \cdots \\ e_{NnN} \end{array} \end{array}
\begin{array}{c} \begin{array}{cccc} C_1 & C_2 & \cdots & C_N \\ e_{11} \cdots e_{1n1} & e_{21} \cdots e_{2n2} & \cdots & e_{N1} \cdots e_{NnN} \end{array} \\
\begin{bmatrix} W_{11} & W_{12} & \cdots & W_{1N} \\ W_{11} & W_{12} & \cdots & W_{1N} \\ \cdots & \cdots & \cdots & \cdots \\ W_{11} & W_{12} & \cdots & W_{1N} \end{bmatrix} \end{array}
$$

Figure 4.3 Supermatrix

obtained, priority in reference to the general objective of the decisional model. These vectors form the matrix weight of the clusters.

More specifically, the operation of weighting takes place through a multiplication. The first entry of the vector is, in fact, multiplied by all the elements in the first block of the column. The second entry of the vector is, in fact, multiplied by all the elements in the second block of the column, and so on. The result of this operation is the so-called weighed supermatrix. Importantly, this supermatix ends ups being stochastic – that is, the sum of the values contained in each column is one.

Once the weighted supermatrix is obtained, it is possible to work on it more in order to obtain the global priorities of all the elements that are part of the decisional model. These priorities are contained in a matrix called the limit supermatrix. In fact, multiplying the weighed supermatrix by itself a number of times that tends towards infinity, as in equation 3, we can obtain the so-called limit supermatrix – that is the supermatrix whose columns contain the vector of the priorities of the elements of the analysis.

$$\lim_{k \to \infty} W^k \quad (3)$$

According to the Perron-Frobenius theorem, in fact, a W matrix with finite elements has a single principal autovector and the equation (3) represents the matrix having the circled autovector as its column.

Finally, we should remember that the ANP model is structured according to the BOCR design. Thus it is necessary to develop the three so-called supermatrices (initial, weighted and limit) for each of the four sub-networks. In this way, four series of weights are obtained for the alternatives (one for benefits, one for costs, one for opportunities and one for risks).

In order to arrive at a single ranking of complex priorities, the formulas in Table 4.3 enable users to integrate the results coming from the development of the single sub-networks, through the aggregation of the results and the final synthesis of the priorities of the alternatives (Saaty and Vargas, 2006).

Table 4.3 Combinatory formulas for the aggregation of the results of the structured-network model

Combinatory formulas	
Additive (or Reciprocal)	B + O + 1/C + 1/R
Probabilistic additive	B + O + (1 − C) + (1 − R)
Subtractive	B + O − C − R
Multiplicative	BO/CR

In operational terms, the proposed formulas indicate the methods for combining the priority vectors resulting from the various sub-networks in order to reach the final priority vector.

As should be emphasized, the first-place alternative in the ranking of the benefits and opportunities sub-networks is the alternative that presents the most benefits and opportunities and therefore the most preferable one. On the contrary, the first-place alternative in the ranking of the costs and risks sub-networks is the alternative that presents the highest costs and risks and is therefore the least preferable. This explains why costs and risks always appear with minus signs or with their reciprocal values in the costs and risks combinatory formulas. In particular, in the multiplicative formula, the values deriving from the benefits and opportunities sub-networks are multiplied by the reciprocal values of the sub-network costs and risks. In this way, the most costly and risky alternative assumes the lowest priorities.

In the following section, the methodological steps above are applied to the TIE methodology experimented in Trentino. The ANP application was supported by Superdecisions software, adopting a structured BOCR model.

4.3 Application of the ANP Model in Trentino

TIE methodology for the management of retail development in Trentino proposes an alternative approach (see Chapter 1) to the standard established strategic evaluation procedures in regional planning (Gauthier et al., 2011; Nilsson and Dalkmann, 2001; Vicente and Partidario, 2006).

To organize the structure of the evaluation, TIE adopts and adapts some of the elements of the MCA, ANP and SWOT analysis methods, making a number of adjustments and changes. In particular, the application of the analytical procedure based on multiple criteria and modelled after the ANP network was developed in the context of TIE in order to define the priorities of the alternate criteria as well as of the 'what-if scenarios' for the settlement of Gross Leasable Areas (GLAs).

Starting out from the SWOT analysis, it was possible to identify the critical elements and the elements of success in relation to the identification of scenarios for the establishment of GLAs (Retail, Marketing, and Landscape) (see Chapter 3). Such elements were later grouped into 'clusters', with reference their specific natures and the characteristics they represent. Therefore relationships of dependence and feedback were identified, the relationships that connect the elements of the system, as well and their respective clusters.

As outlined in the methodology section, such relationships can be relationships of reciprocal dependence (in the sense that one depends on the other), or of unilateral dependence (in the sense that only one depends on the other). At other times, the relationships are only those of influence or of cause and effect. The identification of such relationships brought about a definition of a network system for each system (or sub-network) of the control hierarchy defined by the SWOT analysis (corresponding to the BOCR model).

The following are the results that can be expected from an operational use of such a decisional tool (on the municipal and provincial scales) by the stakeholders involved in the decision-making process:

- the definition of provincial-scale strategies for the territorial enhancement through policies that take into account certain requirements for the areas for the new GLAs – accessibility, the availability of the surface area, no-longer-used areas, low environmental constraints and so on;
- guidelines for orienting decisions for the settlement of local retail areas, defining the priorities indispensable for the social and environmental integration and sustainability of the projected proposals, which are subject to competitive supply.

The following sections will briefly illustrate the construction of an ANP network model for the choice of territorial retail scenarios in Trentino and the application of the model as a decisional support tool on two scales – macro and micro.

4.3.1 The Definition of the ANP Evaluation Model for the Construction of Territorial Scenarios

As already explained in Chapter 1, the structure of the TIE process relates to two interwoven scales of evaluation which correspond to two levels of territorial governance, each with its own precise mandate, purposes and resources: the macro-territorial level of evaluation (VCs), and the micro-territorial level of evaluation (cities). Once the foreseeable scenarios in Trentino were defined, two decisional models were constructed for each territorial scale – micro and macro. Each model was composed of a main network and secondary networks (sub-networks).

The main network was developed starting from a SWOT analysis (corresponding to the BOCR approach illustrated in Chapter 4.2.1), where:

- Benefits and costs (B–C) become the control criteria for a complex network model at the micro scale;
- Opportunities and risks (O–R) are the control criteria at the macro scale.

These criteria are weighted on the basis of two spatial conditions (see Chapter 3): 'accessibility' and 'functional integration' (see Figure 4.4 for the micro scale). This allows control criteria to be prioritized (or weighted) in relation to the strategic requirements previously identified by the Provincial Territorial Plan (PTP).

The development of the sub-networks required that these factors be identified:

- A cluster of alternatives, made up of the three scenarios;
- A variable number of indicators grouped in homogeneous clusters in relation to the SWOT analysis and filtered through the use of related control criteria (that is, B–C at the micro scale and O–R at the macro scale). The clusters correspond to the five strategic evaluation. The number of clusters varies according to the SWOT analysis (see figures 4.5 and 4.6 for the macro and the micro scale);
- The relationships among the sub-network nodes establishing the consequent relationships among clusters. These relationships were identified by a focus group exercise organized in October 2011 that involved all the research team members.

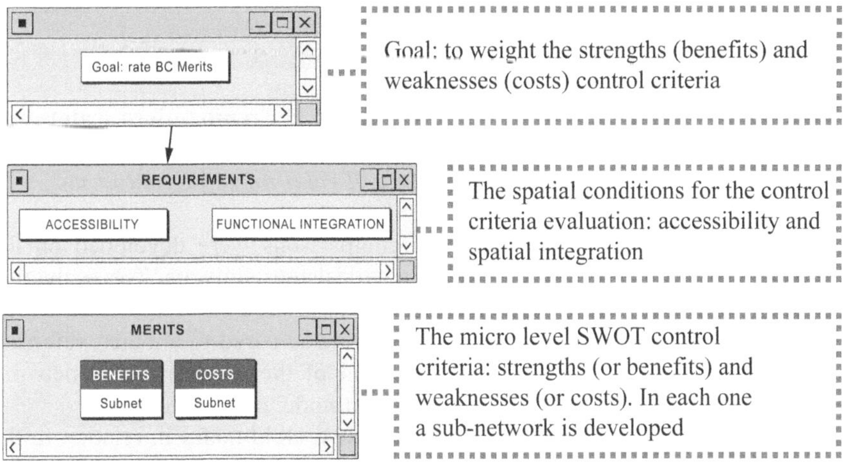

Figure 4.4 Main network at the micro scale

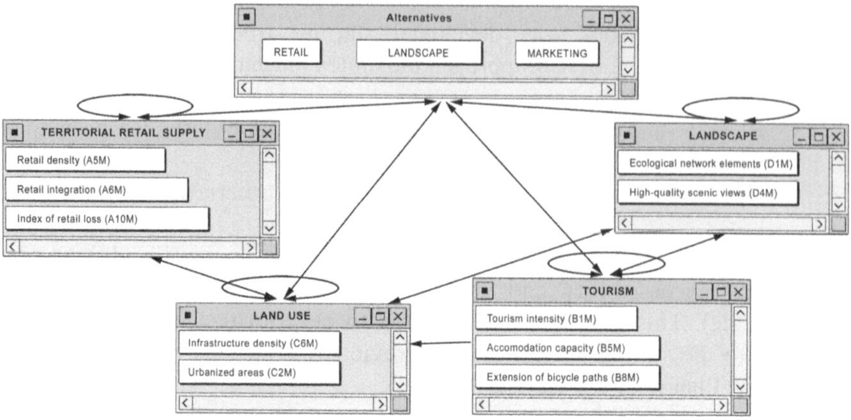

Figure 4.5 Sample of the risks sub-network, Val d'Adige VC (macro scale)

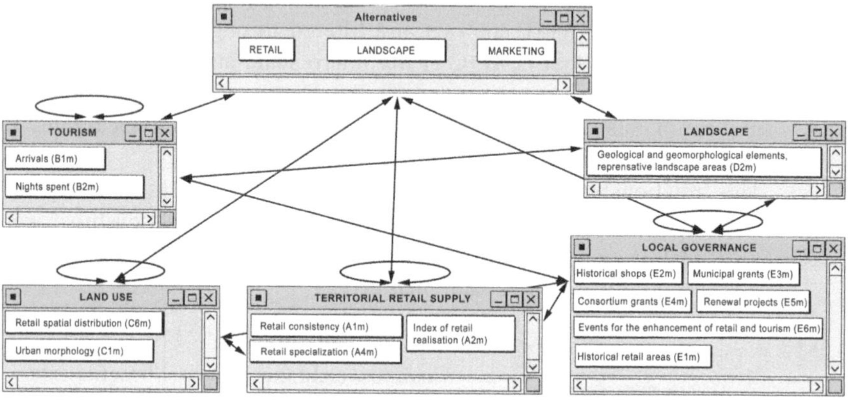

Figure 4.6 Sample of the benefits sub-network, Trento (micro scale)

4.3.2 Synthesis of the Priorities in the Case of Val d'Adige VC and Trento

For each sub-network, several paired comparisons were developed on the basis of the relations identified, using Superdecisions software. The evaluation process required that the various stakeholders participate in the focus group exercise. Each node (indicator) and cluster is compared to another, referring to a specific goal, which is another element of the network. Each network element could switch roles as either a 'parent node' or a 'child node'.

A binary relation is established between two child nodes in relation to the parent node (or comparison criterion). The possible question could be: How important is node A compared to node B, considering C as a goal? The evaluation process uses Saaty's Fundamental Scale (see Table 4.1) to assess the priority of the elements.

The expected results from the paired comparisons are:

- the importance (or weight) of the clusters as TIE strategic themes in relation to the relative control criterion;
- indicator priorities (as actions) in relation to the relative control criterion;
- alternatives or spatial scenarios sorted by priority (weight) at the sub-network level.

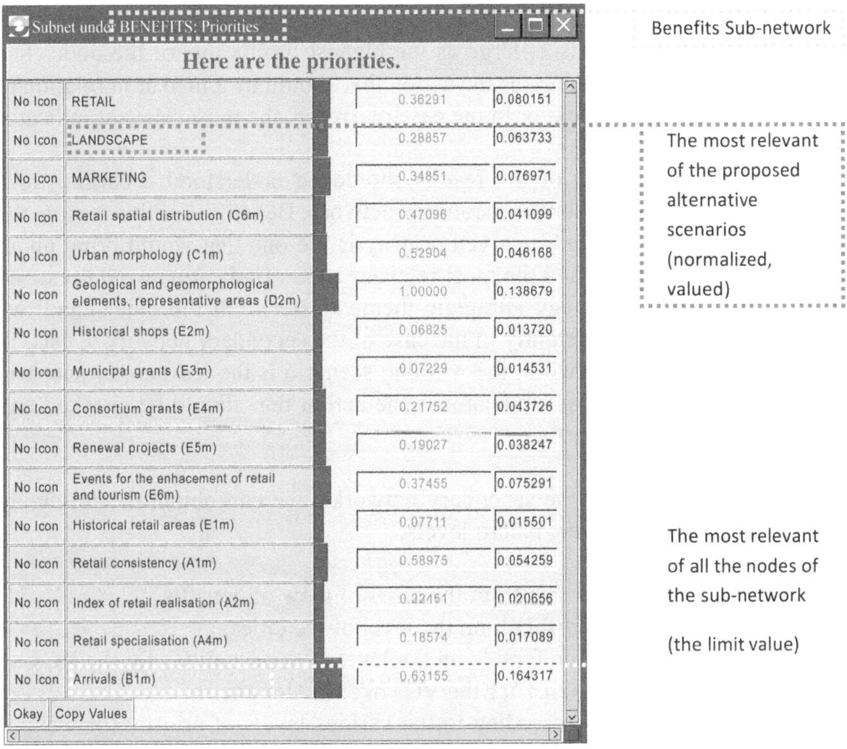

Figure 4.7 Priorities of the sub-network benefits, Trento

Figure 4.7 is a graphic representation of these priorities in reference to the case study of the sub-network Benefits of the city of Trento (micro level):

- The weight of the element normalized in relation to the cluster it belongs to. This value (in the fourth column) indicates the level on the scale of priorities that the element has in relation to the other nodes that compose the same cluster. The sum of the normalized values of the nodes belonging to its same cluster should be equal to one. Hence, in the cases where there is only one indicator, as happens for D4m, its value is one. This allows us to draw up an authentic ranking of the actions (nodes/indicators) – from the most important to the least – to take in order to activate the strategy that was referred to and expressed in the mission of the cluster – that are its strategic theme;

- The limit weight of the element. This value expresses the element's weight in reference to the sum total of all the other elements that make up the network (in the fifth column). The sum of all the limit values turns out to be equal to one. The blue histogram, highlighted in the figure, refers to that value and is as large as the highest limit value of the node. The 'winning' node expresses the action that should be aimed at in relation to the prevailing scenario. In the case of Trento, the most relevant indicator is B1m – 'Arrivals';
- The interpretation of the results should be understood according to a positive meaning for the secondary network, Benefits. For this reason, the scenario that should be given priority is the one that would bring about the most benefits and the most relevant action (indicator) would respond to a winning strategy (strategic theme/cluster). On the other hand, we should reason differently in the case of the secondary network, Costs. In this secondary network, the winning scenario is the least costly scenario and the predominant indicator is the action that should be checked and re-checked the most.

From this evaluation of the secondary networks, we can obtain the following results on the micro and the macro levels:

- the scenario preferable from the point of view of benefits;
- the scenario most difficult on the level of the cities;
- the scenario most preferable on the level of opportunities for the VCs;
- the scenario most risky for the VCs over the long term;
- indicators ranked according to the various levels of priority (or weight) from the point of view of benefits, costs, opportunities and risks.

The most relevant spatial scenario at both the micro and the macro level was obtained through the following synthesis formula (formula 1 and formula 2), corresponding to the additional one of Table 4.3:

$NormalNet(B) * $SmartAlt(B) + $NormalNet(C) * $SmartAlt(C) (1)
$NormalNet(O) * $SmartAlt(O) + $NormalNet(R) * $SmartAlt(R) (2)

where:

$NormalNet(node name) refers to the normalized value of the node in relation to the current network;

$SmartAlt(node name) refers to the best alternative in relation to the current network.

As recommended by Adams and Saaty (2003), influence was treated consistently – how the parent influences the children, or vice versa – but the flow direction was kept the same throughout the network and throughout the model. The Superdecisions software used for making comparisons and deriving priorities has allowed us to discover comparison coherence levels in real time.

4.3.3 Discussion of the Results

The overall priorities for the case study are illustrated in Table 4.4 and Table 4.5. These show the final priorities for the macro scale of Val d'Adige VC and for the micro scale of Trento, where the indicators are ordered starting from the most important, which is also the most interrelated.

Table 4.4 Val d'Adige VC ANP overall priorities (macro scale)

Subnet	Scenarios		Actions		Nor	Lim
Criteria	Accessibility is 0.25		Integration is 0.75			
Weights	Risks are 0.38		Opportunities are 0.62			
Opportunities	Retail	0.50	A9M		1.00	0.25
	Marketing	0.27	A1M		0.66	0.25
	Landscape	0.23	B8M		0.33	0.12
Risks			D1M		0.72	0.19
	Retail	0.50	C2M		0.50	0.14
	Marketing	0.26	C6M		0.49	0.13
	Landscape	0.24	D4M		0.27	0.07
			B8M		0.49	0.06
Aggregation of the results			Retail	+0.50		
			Marketing	+0.26		
			Landscape	+0.24		

Note: A9M – 'Index of retail choice'; A1M – 'Retail consistency'; B8M – 'Extension of bicycle paths; D1M – 'Ecological network elements'; C2M – 'Index of urban growth'; C6M – 'Infrastructure density'; D4M – 'High quality scenic views'.

Source: Author's elaboration.

On the macro scale, there are more opportunities for the Val d'Adige VC than there are risks. In relation to the opportunities, Scenario 1 turned out to have the highest weight. The reason is that the indicators that make reference to A1M – 'Retail consistency' and to A9M – 'Index of retail choice' were valued as highly influential. In relation to risks, again Scenario 1 takes on the highest weight because the criteria that had to do purely with the retail sector were not considered risky. Rather, they were considered to be nodes that involved the landscape and surface area consumption.

Table 4.5 Trento ANP overall priorities (micro scale)

Subnet	Scenarios		Actions		Nor	Lim
CRITERIA	Accessibility is 0.83		Integration is 0.17			
WEIGHTS	Benefits are 0.73		Costs are 0.27			
Benefits	Retail	0.36	B1m		0.63	0.16
	Marketing	0.34	D2m		1.00	0.13
	Landscape	0.28	B2m		0.37	0.09
Costs			E6m		0.37	0.07
	Marketing	0.40	A1m		0.60	0.05
	Retail	0.35	D1m		0.74	0.22
	Landscape	0.25	C2m		0.58	0.16
			A3m		0.26	0.07
Aggregation of the results			Retail	+0.37		
			Marketing	+0.33		
			Landscape	+0.30		

Note: B1m – 'Arrivals'; D2m – 'Geological and geomorphological elements, representative landscape areas'; B2m – 'Nights spent'; E6m – 'Events for the enhancement of retail and tourism'; A1m – 'Retail consistency'; D1m – 'Ecological network elements'; C2m – 'Index of urban growth'; A3m – 'Retail typology'.
Source: Author's elaboration.

The aggregation of the results according to the additive formula (2) outlined above leads us to emphasize that the winning long-term scenario is Scenario 1. The reason is that this scenario is marked by the high quality of the existing supply, which can be integrated with new expansion into specialized fields. Instead, Scenario 2 and Scenario 3 turned out to be of equal value. This is an indication of the need to increase the level of the flow of tourists in relation to the excellent bicycle-path infrastructure, where there is a low level of dynamism in the tourist sector.

On the micro scale, accessibility turned out of have a high weight because Trento is crossed by the Brenner motorway and by a significant number of efficient roads with heavy traffic, roads that link the Trento with Pergine Valsugana (route ss. 47), Rovereto (route ss. 12), Arco and Riva del Garda (route ss. 45bis). In this context, territorial integration can play a helping role in the construction of evolutionary scenarios directed at historical-cultural resources and at the tourism of cultures and fairs, which would serve as the keys for the promotion of a winning territorial strategy. In relation to benefits, the three scenarios turn out to have the same influence, having a weight that is relatively higher than that of Scenario 1. Instead, in reference to costs, Scenario 2 has the highest weight. The reason is that, despite a good level of tourist arrivals and stays, the system of tourism needs to be integrated with a specialized commercial system, directing itself more towards local projecting initiatives.

The aggregation of the results according to the additive formula (1) shows that, to date, Scenario 1 turns out to be the most developed scenario. The reason is that the retail supply in Trento is a strong point at micro and macro scale in terms of consistency and specialization. Instead, Scenario 2, 'Marketing' and Scenario 3, 'Landscape' are marked by the high influence of the historical-cultural resources and a good local-governance drive. However, these need to be associated with a strategy that integrates retail and tourism from the perspective of creating complementary relationships with other nearby cities, like Rovereto in Vallagarina VC.

4.4 The Contribution on ANP in TIE Methodology

In the TIE methodology, the ANP method was developed at both the macro scale (provincial) and the micro scale (municipal), comparing the indicators' priorities to the relative control criterion for a better definition of territorial scenarios. In particular, the ANP method has enabled us to select the most relevant indicators defining the spatial scenario, indicators that are obtained through an appropriate synthesis formula.

This study demonstrates that the ANP application represents an operational tool for policy makers' useful for supporting decision-making in spatial planning. In particular, the application carried out on Trento Province, at both micro and macro scale, has allowed us to identify the most appropriate scenario for retail planning. In addition, the application has allowed us to define the final indicator-framework, the framework useful for the description of shared decision-making problems and representative of all the dynamics.

Thanks to the network-modelling representation of the problem and the adoption of SWOT analysis for the structuring of the process,, the ANP methodology has been able to support:

- the definition of strategy for land-use rationalization, increasing territorial value through land marketing policies and considering some crucial requirements;

- the definition of the indicator priority for the integration and social and environmental sustainability of the design/planning proposals.

The outcome of this TIE process, as supported by an MCA and ANP approach, has been recently converted into local legislation (see Chapter 6).

References

Adams, W.J. and Saaty, R., 2003. *Super Decisions Software Guide.* [Online] Available at: http://www.superdecisions.com [Accessed 15 January 2013].

Bottero, M. and Ferretti, V., 2011. An analytic network process-based approach for location problems: The case of a waste incinerator plant in the province of Torino (Italy). *Journal of Multicriteria Decision Analysis*, 17(3–4): 63–84.

Bottero, M., Lami, I. and Lombardi, P., 2008. *Analytic Network Process: La valutazione di scenari di trasformazione urbana e territoriale.* Firenze: Alinea Editrice.

Brandon, P.S. and Lombardi, P., 2011. *Evaluating Sustainable Development in the Built Environment.* Oxford: Wiley-Blackwell.

Deakin, M. and Lombardi, P., 2005. A Directory of Environmental Assessment Methods. In: S. Curwell, M. Deakin and M. Symes (eds), 2005. *Sustainable Urban Development: The Framework, Protocols and Environmental Assessment Methods.* London: Routledge, pp. 175–92.

Farkasovsky, M. and Greda, A., 2006. Outsourcing a Firm's Application Development Grouping. In: T.L. Saaty and L.G. Vargas (eds), 2006. *Decision Making with the Analytic Network Process.* New York: Springer Science.

Figueira, J., Greco, S. and Ehrgott, M. (eds), 2005. *Multiple Criteria Decision Analysis: State of Art Surveys.* New York: Springer.

Gauthier, M., Simard, L. and Waaub, J.P., 2011. Public participation in strategic environmental assessment (SEA): Critical review and the Quebec (Canada) approach. *Environmental Impact Assessment Review*, 31(1): 48–60.

Gomez-Navarro, T., Garcia-Melon, M., Acuna-Dutra, S. and Diaz-Martin, D., 2009. An environmental pressure index proposal for urban development planning based on the analytic network process. *Environmental Impact Assessment Review*, 29(5): 319–29.

Greco, S., Matarazzo, B. and Slowinski, R. 2001. Rough sets theory for multicriteria decision analysis. *European Journal of Operational Research*, 129: 1–47.

Lombardi, P., 2007. The Analytic Hierarchy Process. In: M. Deakin, G. Mitchell, P. Nijkamp and R. Vreeker (eds), 2007. *Sustainable Urban Development: The Environmental Assessment Methods*, vol. 2. Abingdon: Routledge, pp. 209–22.

Lombardi, P., 2009. Evaluation of sustainable urban redevelopment scenarios. *Proceedings of the Institution of Civil Engineers: Urban Design and Planning*, 162(4): 179–86.

Lombardi, P., 2010. Strategic management plan evaluation of a river basin district. *Management of Environmental Quality*, 21(1): 32–44.

Miller, D. and Patassini, D. (eds), 2005. *Beyond Benefit Cost Analysis*. Aldershot: Ashgate.

Munda, G., 2005. Multiple Criteria Decision Analysis and Sustainable Development. In: J. Figueira, S. Greco and M. Ehrgott (eds), *Multiple Criteria Decision Analysis: State of the Art Surveys*. New York: Springer, pp. 953–86.

Nijkamp, P., Rietveld, P. and Voogd, H., 1990. *Multicriteria Evaluation in Physical Planning*. Amsterdam: Elsevier.

Nilsson, M. and Dalkmann, H., 2001. Decision making and strategic environmental assessment. *Journal of Environmental Assessment and Policy Management*, 3(3): 305–27.`

Roy, B. and Bouyssou, D., 1995. *Aide multicritère à la décision: Méthodes et case*. Paris: Economica.

Saaty, T.L., 1980. *The Analytic Hierarchy Process*. New York: McGraw Hill.

Saaty T.L., 1990. *Decision-making for Leaders*. Pittsburgh, PA: RWS Publications.

Saaty, T.L., 1992. *Multicriteria Decision Making: The Analytic Hierarchy Process*, Pittsburgh, PA: RWS Publications.

Saaty, T.L., 1993. What is a relative measurement? The ratio scale phantom. *Mathematical and Computer Modelling*, 17(4 5): 1–12.

Saaty, T.L., 1994. *Fundamentals of Decision Making and Priority Theory with the Analytic Hierarchy Process*. Pittsburgh, PA: RWS Publications.

Saaty, T.L., 1995. *Decision Making for Leaders*. Pittsburgh, PA: RWS Publications.

Saaty, T.L., 1996a. *Decision-making with Dependence and Feed-back: The Analytic Hierarchy Process*. Pittsburgh, PA: RWS Publications.

Saaty, T.L., 1996b. *The Analytic Network Process*. Pittsburgh, PA: RWS Publications.

Saaty, T.L. 2000. *Fundamentals of Decision-making and Priority Theory with the Analytic Hierarchy Process*. Pittsburgh, PA: RWS Publications.

Saaty, T.L., 2001. *The Analytic Network Process: Decision Making with Dependence and Feedback*. Pittsburgh, PA: RWS Publications.

Saaty, T.L., 2005. *Theory and Applications of the Analytic Network Process: Decision Making with Benefits, Opportunities, Costs, and Risks*. Pittsburgh, PA: RWS Publications.

Saaty, T.L. and Forman, E., 1993. *The Hierarchon*. Pittsburgh, PA: RWS Publications.

Saaty, T.L. and Ozdemir, M.S., 2008. *The Encyclicon: A Dictionary of Complex Decisions Using the Analytic Network Process*. Pittsburgh, PA: RWS Publications.

Saaty, T.L. and Vargas, L.G., 1994. *Decision Making in Economic, Political, Social and Technological Environments*. Pittsburgh, PA: RWS Publications.

Saaty, T.L. and Vargas, L.G., 2006. *Decision Making with the Analytic Network Process*. Springer: New York.

Vicente, G. and Partidario, M.R., 2006. SEA enhancing communication for better environmental decisions. *Environmental Impact Assessment Review*, 26(8): 696–706.

Voogd, H., 1983. *Multicriteria Evaluation for Urban and Regional Planning*. London: Pion.

Wijnmalen, D.J.D., 2007. Analysis of benefits, opportunities, costs, and risks (BOCR) with the AHP–ANP: A critical validation. *Mathematical and Computer Modelling*, 46(7/8): 892–905.

Chapter 5
A Mathematical Model for Territorial Integrated Evaluation

Roberto Monaco

5.1 Aims and Benefits of the Model

This chapter proposes a mathematical model that can evaluate mobility – that is, the flows of people inside a multi-pole territorial system. In this case, the poles are the Gross Leasable Areas (GLAs) in the 16 Valley Communities (VCs) in Trentino. The mobility of people is influenced by the intrinsic attractiveness of a pole, by its distance from the other poles, and by its carrying capacity, a factor that takes into account the effects of overcrowding. Attractiveness, distance and carrying capacity are factors that can be quantified suitably, becoming the parameters of the model. Attractiveness is determined according to an accurate multivariate statistical method over a suitable number of indicators specifying the characteristics of each pole. The final section presents a test of a numerical simulation of the behaviour of the model, which seeks to estimate the potential flows of people among the GLAs in some of Trentino's VCs.

The mathematical model takes up some of the ideas developed in the 1980s. Back then, mathematical models were proposed that could represent the mobility of people inside an area, where several centres of attraction ('poles') are present. In this regard, the paper by Monaco and Rabino (1984) takes Lotka-Volterra models (Murray, 2002) into consideration. These are models that fit very well into our investigation because they exhibit the existence of states of equilibrium, thanks to logistic terms that assure retroaction effects. Hence the presence of the stability of equilibrium solutions assures us that we can check the controllability of mobility flows inside the system. In their most general form, these models may handle interactions among residents in the poles, both their cooperation and their competition. They may also handle the flows of emigration and immigration in and out of an area. Nevertheless this chapter will treat neither the interactions among residents nor emigration–immigration flows.

Hansen (1959) and Wilson (1974) emphasize that the accessibility of each pole plays an important role. They evaluated a pole's accessibility not only in terms of economic, social and cultural attractiveness but also in terms of its distance from other poles. As Martinetti (1993) and Carrothers (1956) have pointed out, attractiveness may strictly depend on the facilities in the area itself. Hence we need to quantify the intrinsic attractiveness of each pole in order to determine flows of people inside the system. Only this will allow us to derive a mathematical model. Obviously, attractiveness is a complex quantity that depends on various components. Thus we need statistical tools to determine attractiveness.

Here, we will determine attractiveness by starting out with the non-aggregate data of each pole, consisting in a GLA. This data will be analysed by means of a multivariate statistical technique in order to determine the main components of the data, components that are the intrinsic characteristics (indicators) of the area itself. Finally, these indicators will be put together suitably, and this composition of indicators is the factor that determines the attractiveness of a pole. This is the procedure that we followed in other studies (see Monaco and Servente, 2006). The following section will present a suitable mathematical model for an N-pole system, including the spatial scale of an area in terms of the distances of the poles from each other, determined by its existing road network.

The model proposed in this section belongs to a general class of models described more recently in Monaco and Servente (2006). This model was adapted to the particular needs of this application of Territorial Integrated Evaluation in this book, taking into account only the parameters linked to the poles' attractiveness, accessibility and carrying capacities. The carrying capacity is a factor that is determined by the so-called crowding indexes, which normally can be deduced from the security standards of the GLAs. In fact, carrying capacity also can be seen as a kind of threshold of comfort. Namely, when the number of customers crosses this threshold, customers are no longer motivated to visit a GLA. The results that this model yields are relative results. We should not dwell on the absolute values of the numerical results. Rather, we should compare the numerical results obtained by the various GLAs with one another, paying special attention to the general trend of the system during its evolution.

5.2 The Mathematical Model

Let us consider a region characterized by N poles (in our case the number N of VCs), each VC linked with the others by a regular road network. We may assume that we know the distances d_{ij} between each pair of poles, where i and j are the subscripts identifying the i-th and j-th pole, $i,j=1 \dots N$. In addition, let us indicate by $p_i(t)$ the number of people who are present in the i-th pole at time t. Obviously, we may assume that the number $p_{i0}=p_i(0)$ of people who are present at time $t=0$ in the i-th pole is known. Furthermore, each pole is characterized by the index of attractiveness indicated by A_i – for people present in the pole i – and by B_{ij} – for people present in the poles $j \neq i$ who are attracted to move to the pole i. The indexes of attractiveness are determined through a multivariate statistical analysis, which will be explained in the next section. As stated previously, we will disregard terms that take into account interactions between people who are present in different poles and immigration/emigration processes due to people coming in or out the region under consideration. Thus, in accordance with the modelling proposed in Monaco and Servente (2006), this model takes the form:

$$p'_i = A_i\, p_i \left(1 - \frac{p_i}{s_i}\right) + \sum_{j=1, j\neq i}^{N} B_{ij}\, p_j \quad (1)$$

In equation (1), p'_i is the time derivative of p_i. The first term in the right-hand-side takes into account, with a logistic expression (Murray, 2002), the attractiveness of the i-th pole for people already present in that pole. The second term takes into account the attractiveness of the i-th pole on people present in the other j poles. The logistic term of equation (1) depends on the carrying capacity of the pole i, which is given by the expression in parentheses which is affected by the threshold s_i indicating the maximum number of people beyond which the pole loses its quality of comfort. In short, it becomes overcrowded. Equation (1) represents a set of N ordinary differential equations, joined to the initial data p_{i0}.

We can now specify the attractiveness B_{ij} better. As stated in Chapter 5.1, the attractiveness of two different poles to each other also depends on their distance. Therefore we can claim that the parameter B_{ij} is proportional to the actual 'absolute' attractiveness A_i, determined by the intrinsic properties of the pole i, through a suitable function f of the mutual distance d_{ij}, that is,

$$B_{ij} = f(d_{ij})\, A_i. \quad (2)$$

In the last expression it is assumed, however, that each pole has a single value of attractiveness for any individual present in the other poles. This is a consequence of the fact that the indexes of attractiveness are related only to the intrinsic properties of each pole i. As discussed in the references (Carrothers, 1956; Hansen, 1959; Wilson, 1974), the function f must take into account the fact that distance between two poles is an impediment to accessibility. In other words, the greater the distance, the lesser the attractiveness. This hypothesis is reasonable if one considers that the mobility between poles far away from each other is only marginal.

Some authors, for instance Wilson (1974), suppose that the function f decays exponentially as distance increases. However, we think that it is preferable to assume that it decays algebraically. Thus, as in Monaco and Servente (2006), we propose the following expression for the function f

$$f(d_{ij}) = \frac{1}{1 + \beta\, d_{ij}^2} \quad (3)$$

where β is a suitable factor identifying the territorial scale under investigation.

As in Monaco and Servente (2006), the determination of the factor β is performed in the following way: approximate expression (3) with a McLaurin expansion of order two, assuming that the function f vanishes when computed for $d_{ij} = d_M$, where d_M is the maximum value of the distances d_{ij} for any value of i and j in the road network. Such a McLaurin expansion provides the following expression for β

$$\beta = \frac{1}{d_M^2} \quad (4)$$

Let us note that, according to this hypothesis, d_M takes on the meaning of the characteristic scale of the territorial region under investigation. By introducing the expressions (2), (3) and (4) in the equations (1), we get the final expression of our model equations, that is:

$$p'_i = A_i\, p_i \left(1 - \frac{p_i}{s_i}\right) + A_i \sum_{j=1, j\neq i}^{N} \left[1 - \left(\frac{d_{ij}}{d_M}\right)^2\right] p_j \quad (5)$$

Thanks to the logistic term present in these equations that plays the role of retroaction, such a model assures equilibrium solutions that are asymptotically stable. The asymptotic trend to these equilibrium states can simulate the mobility of people in our current system of poles as well as their presence in the GLAs themselves. In the last section of this chapter, we will present a case study and show how the results of the model should be interpreted.

5.3 The Application of the Model in Trentino

The system of the VCs, each characterized by the attractiveness of its GLA, can be represented by a geometrical graph, where the nodes are the GLAs of the VCs and the arches the roads in Trentino connecting the VCs with one another. Nodes and arches can be quantified according to the attractiveness of their GLAs and according to their distances from each other, determined by the road network.

Such a graph is the geometrical reference that can be used to implement the mathematical model that can evaluate the mobility flows of people throughout the GLA system. As already discussed, mobility depends on the number of people present in the VC, on its attractiveness, on the carrying capacity of each GLA and on the accessibility of each GLA, determined by the distance matrix. Therefore we need to recover the following data in order to implement the mathematical model:

- number of people resident in each VC;
- mean values of seasonal tourist presences, a value to be added those of the residents;

- current distance network between the GLAs over the roads;
- size of each GLA in terms of its current square meters;
- carrying capacity of each GLA, evaluated in terms of the crowding indexes.

Further data must be available in order to determine attractiveness through a suitable multivariate statistical analysis. In fact, attractiveness depends on the economic condition of each VC. This is a factor that may be determined by evaluating the correlations between the infrastructural and productive facilities of each GLA in order to check whether the localisation of retail activities is favourable or not. Thus attractiveness must be understood as a macro indicator for each *i*-th VC, which is a synthesis of two conditions: accessibility (indicator X_i) and functional integration (indicator Y_i).

The condition accessibility for each VC is a synthesis of these factors:

- the primary road system (index C_i);
- the secondary road system (index D_i);
- the presence of a motorway and railway hub (index E_i);
- the VCs territorial surface (index F_i).

Such a synthesis allows us to determine the size of the territorial accessibility.

The condition functional integration for each VC is a synthesis of these factors:

- retail consistency (index G_i);
- tourism nights spent (index H_i).

Particularly, the indexes above have been explicitly determined through the following normalized ratios (all numbers between 0 and 1):

- Ci = (km of VC motorways)/(km of Province motorways and provincial roads);
- Di = (km of VC provincial roads)/(km of Province motorways and provincial roads);
- Ei = (VC territorial surface)/(Province territorial surface);
- Fi = presence of a motorway and railway hub (yes/no 0/1);
- Gi = (VC GLA surface)/(Province GLAs surface);
- Hi = (VCs night spent)/(Province nights spent).

This data was handled for all 16 VCs by a software program performing principal-component analysis (Johnson and Wichern, 2008), which obtained the following weighted average values for the indicators X_i and Y_i:

$$X_i = (2C_i + D_i + E_i + F_i)/5$$
$$Y_i = 0.7G_i + 0.3H_i.$$

Finally we have combined the values of the indicators X_i and Y_i in the following way to obtain the index of attractiveness A_i as a normalized indicator:

$$A_i = 0.6X_i + 0.4Y_i.$$

The values of A_i for all the VCs are given in Table 5.1.

Table 5.1 Index of attractiveness (A_i) of the VCs

Valley Community	*Ai*	Valley Community	*Ai*
Val d'Adige	1.00	Val di Non	0.38
Vallagarina	0.84	Val di Fiemme	0.37
Rotaliana-Könisberg	0.60	Valsugana-Tesino	0.33
Alto Garda-Ledro	0.58	Paganella	0.32
Giudicarie	0.48	Primiero	0.30
Alta Valsugana-Bersntol	0.46	Altopiani Cimbri	0.29
Val di Fassa	0.42	Valle di Cembra	0.25

Source: Author's elaboration.

The following section traces how we set up the evaluation of the flows of mobility of customers among the VCs, focusing only on the four VCs that yield the highest values of attractiveness (that is Val d'Adige, Vallagarina, Rotaliana-Könisberg and Alto Garda-Ledro). In contrast to all the other VCs, these are the valleys that are strongly characterized by a solid retail system, as illustrated in the application of Territorial Integrated Evaluation (TIE) (see Chapter 3, Theme A: 'Territorial Retail Supply'). In addition, compared to the other three VCs, Alto Garda-Ledro VC is characterized not only by its distinctive primary theme – retail – but also by its secondary tendency towards tourism and landscape – themes more characteristic of the 10 VCs with lower values of attractiveness (see Chapter 3.1). These are the reasons why we have restricted our analysis of mobility to these four VCs only, in view of these results and the goals of this chapter. Furthermore, an analysis was made to determine the dominant and prospective scenarios of these VCs. In fact, the determination of the dominant scenario is important since it shows the framework of the dynamic development of each VC, as evidenced from the analysis of the risk/opportunity factors. The results of such an analysis show that Val d'Adige, Vallagarina and Rotaliana-Könisberg are characterized by the dominant Scenario 1 – Retail, whereas Alto Garda-Ledro is characterize by the Scenario 3 Landscape (see Chapter 3.2). The next section we will illustrate the role played by the coupling of retail activities and tourism the themes that characterize Alto Garda-Ledro.

5.4 Discussion of the Results

By using the mathematical model defined in Chapter 5.2, here we present a simulation performed on the four VCs, with the aim of validating the model itself. The results must be considered preliminary and not completely systematic since they do not take into account any interactions with the other twelve VCs.

The VCs chosen are Val d'Adige, Vallagarina, Rotaliana-Könisberg and Alto Garda-Ledro VCs. This choice was determined by the fact that these four communities present the highest index of attractiveness A_i. In Table 5.2, we give the other data for each VC that is needed to implement the simulation – that is, the number of residents, the mean presences during the summer of tourists and the thresholds of the GLAs, obtained by considering the total surface of the GLA itself multiplied by a security index.

Table 5.2 Number of residents, tourist presences and GLA thresholds in each VC

Valley Community	Residents	Tourist presences	GLA thresholds
Val d'Adige	120,360	245,972	2,537
Vallagarina	88,481	231,467	1,792
Rotaliana-Könisberg	28,903	29,121	682
Alto Garda-Ledro	48,409	2,252,826	1,075

Source: Author's elaboration.

Together with these data, we indicate the distances in kilometres between the GLA of each VC in Table 5.3.

Table 5.3 Distances in km between the GLAs of each VC

Valley Community	Val d'Adige	Vallagarina	Rotaliana-Könisberg	Alto Garda-Ledro
Val d'Adige	—	28	12	36
Vallagarina	28	—	35	24
Rotaliana-Könisberg	12	35	—	42
Alto Garda-Ledro	36	24	42	—

Source: Author's elaboration.

As explained in the previous sections, the model equations (5) determine, through their asymptotical analysis, the steady forecast distributions of individuals inside each VC. Therefore we have compared the distribution of residents (or the distribution of residents plus tourists) and the steady distributions. This is the comparison that allows us to evaluate whether a GLA of each VC is over-sized or under-sized. The results, expressed in percentage, are reported in the table below.

Table 5.4 Comparison between the distribution of residents/ distribution of residents plus tourists and steady distribution in each VC

Valley Community	Steady distribution	Resident distribution	Difference	Resident and tourist distribution	Difference
Val d'Adige	29.0%	42.0%	−13.0%	39.1%	−10.1%
Vallagarina	27.8%	30.9%	− 3.1%	28.9%	−1.1%
Rotaliana-Könisberg	17.6%	10.2%	+7.5%	9.4%	+8.3%
Alto Garda-Ledro	25.6%	16.9%	+8.7%	22.6%	+3.0%

Source: Author's elaboration.

The steady distribution of individuals in the GLAs of each VC is reported in the second column. These percentages show that the communities of Val d'Adige, Vallagarina and Alto Garda-Ledro present a similar distribution of customers – between 25 per cent and 29 per cent – and that Rotaliana-Könisberg is significantly lower – around 17 per cent.

In relation to the distribution of residents (third column), Val d'Adige has a majority of inhabitants (not far from 50 per cent), whereas Alto Garda-Ledro and Rotaliana-Könisberg have a rather small concentration of residents. Thus the fourth column reports the difference between these distributions and the steady ones provided by the model. Such differences indicate that Val d'Adige (−13 per cent) is suffering from a lack of retail attractiveness. Conversely, Rotaliana-Könisberg and Alto Garda-Ledro (+7.5 per cent and +8.7 per cent, respectively) seem to offer an abundance of retail areas.

This kind of analysis is significantly modified if a certain number of tourists are added to the residents. As illustrated in the first table of this section, tourist presences are very high and several times greater than those of residents in three VCs, except for Rotaliana-Könisberg. This is due to the fact that Trentino is one of the most important tourist regions in Italy. The source of these numbers is the determination of the actual numbers of tourist presences during 100 days that include the summer (averaged over several years).

If we include the number of tourist presences, it becomes more complex to analyse the results of the model for three reasons. First, tourists are present over a span of 100 days, but in a random way. Second, the percentage of tourists who are authentic customers is not clear. Third, the data is not balanced because the ratio of tourists to residents differs widely. In Rotaliana-Könisberg it is approximately 1 to 1 and in Alto Garda-Ledro nearly 50–1. Therefore the redistribution of customers can vary significantly in the four VCs and so we should examine the results obtained by the model more carefully. For this reason, we added only 1 per cent of the number of tourists to the number of residents in each VC in the model simulation.

Thus the fifth column of the table above reports the percentage of redistribution of potential customers as the percentage of residents +1 per cent of tourists. Based on the results of the data introduced in the model equations, Alto Garda-Ledro shows a significant increase in potential customers but the three other VCs show a slight decrease. Val d'Adige shows a 3 per cent variation (from −42.0 per cent to −39.1 per cent). Alto Garda-Ledro passes from +8.7 per cent to +3.0 per cent. Rotaliana-Könisberg and Vallagarina do not show any significant variation because the presence of tourists is less important there. According to these results, Val d'Adige still needs an increase in the number of GLA. Alto Garda-Ledro does not show that it has too many GLAs. Its larger number of GLAs may be enough to satisfy customer requests during their stay.

5.5 The Contribution of the Model in TIE Methodology

The results obtained by the application of the mathematical model in Trentino made it possible for us to verify the behaviour of the model itself. The application was restricted to the four VCs with the highest values of attractiveness and the strongest tendency for retail. The model simulations included both the number of residents and of tourist presences in each VC. This number was suggested by a SWOT (Strengths, Weakness, Opportunities, Threats) analysis that defined both the dominant and prospective scenario of each VC (see Chapter 3). Simulations applied to Alto Garda-Ledro VC indicated that it is important to consider the presence of tourists in determining the right size of GLAs. The model functioned rather well and contributed towards a dynamic and systematic evaluation of the territorial retail activities. In the future, it can be applied to the entire system of VCs, where it would allow us to get a general view of the customer flows among the different VCs. Having gained this perspective, planners could strive to obtain the best possible redistribution of the GLAs and support possible new territorial retail scenarios. If the model is implemented all over Trentino, it could indicate potential points of synergy among the different VCs, so that planners could develop territorial retail scenarios that involve not just a single VC but the whole Province.

References

Carrothers, G.A., 1956. A historical review of the gravity and potential concepts of human interaction. *Journal of American Institute of Planners*, 22: 94–102.

Hansen, W.G., 1959. How accessibility shapes land use. *Journal of American Institute of Planners*, 25: 73–6.

Johnson, R.A. and Wichern, D.W., 2008. *Applied Multivariate Statistical Analysis*. Harlow: Pearson.

Martinetti, G., 1993. *Metropoli. La nuova morfologia sociale della città*. Bologna: Edizioni Il Mulino.

Monaco, R. and Rabino, G., 1984. A Stochastic Treatment of a Dynamic Model for an Interacting Cities System. In: X.J.R. Avula, R.E. Kalman, A.I. Liepis and E.Y. Rodin (eds), *Mathematical Modelling in Science and Technology*. New York: Pergamon Press, pp. 326–30.

Monaco, R. and Servente, G., 2006. *Introduzione ai modelli matematici nelle scienze territoriali*. Turin: Celid, Torino.

Murray, J.D., 2002. *Mathematical Biology I: An Introduction*. New York: Springer.

Wilson, A.G., 1974. *Urban and Regional Models in Geography and Planning*. London: Wiley.

Chapter 6

Smart Evaluation: The Agenda Emerging from Experimentation of the Territorial Integrated Evaluation Methodology

Grazia Brunetta

6.1. A Model for Territorial Governance: Towards Open-Scenario Systems

Following the rationale of our experimentation with Territorial Integrated Evaluation (TIE) methodology, a proposal for institutional innovation is in order. This proposal aims to help implement governance processes in the Valley Communities (VCs). There are three spheres of innovation that the proposal has identified:

- systems of scenario – a polycentric spatial arrangement to be defined for the development and enhancement of the Province;
- the design of a management form for the governance of this spatial arrangement (see Chapter 6.2);
- a set of rules to be introduced for the governance and designing of retail territorial development (see Chapter 6.3).

In relation to the first sphere of innovation, the *systems of scenario*, we conducted experimentation that intended to promote the integration of retail territorial policies with regional and landscape-related policies. It also intended to incentivize converging policies involving various levels of planning: provincial, inter-municipal and municipal. The ultimate end was that of encouraging cooperation among the territories/actors/stakeholders that would start to improve the quality of the processes of territorial development. In the interest of this improvement, the results of the TIE provide each VC with a dominant and a prospective territorial retail development scenario, which are based on an evaluation of the current dynamics. As we have seen, the TIE results identify strategies, objectives, actions, and criteria for enhancement, conservation and development to be designed in each VC (see chapters 3 and 4). The following table exemplifies the knowledge framework emerging from the TIE, which is addressed to the Valley Communities for starting the decision-making process about the territorial retail development (Table 6.1).

Table 6.1 The emerging TIE knowledge framework for designing the territorial scenarios

Territories	**Condition**	**Criteria**					**Mid-term scenario**	**Long-term scenario**
Valley Community	*Index of Attractiveness*	*Theme A*	*Theme B*	*Theme C*	*Theme D*	*Theme E*	*Dominant Scenario*	*Prospective Scenario*
Val d'Adige VC	1,00	*					1	2
Vallagarina VC	0,84	*					1	3
Rotaliana-Konisberg VC	0,60	*					1	2
Alto Garda-Ledro VC	0,58	*	*		*		3	
Giudicarie VC	0,48		*		*		3	
Alta Valsugana-Bersntol VC	0,46				*			2
Val di Fassa VC	0,42		*		*		3	
Val di Sole VC	0,40		*		*		3	
Val di Non VC	0,38				*			3
Val di Fiemme VC	0,37				*			3
Valsugana-Tesino VC	0,33				*			2
Paganella VC	0,32		*				2	3
Primiero VC	0,30		*				2	3
Altipiani Cimbri VC	0,29		*				2	3
Val di Cembra VC	0,25							2
Valle dei Laghi VC	0,25				*			2

Source: Author's elaboration.

Note: The leading themes for each VC are marked by an asterisk. The numbers correspond to the type of territorial scenario: 1. Retail: Designing the territorial retail system. 2. Marketing: Designing the integration of retail/tourism/territory. 3. Landscape: Enhancing the landscape identity.

Taking a step beyond this first sphere of strategic lines, we have worked with the aim of making the governance in the VCs actually feasible. Where they exist, we aimed to highlight the potential material and non-material relations among the VCs. We took this last conceptual and operative step in order to help in the designing of governance practices that are as oriented as possible towards gathering the elements of strength and weakness in the current dynamics. These elements of strength and weakness are those that would be exploited in a territorial vision integrated among various 'territories'. From the point of view of the TIE rationality, we highlighted potential factors leading to cooperation of provincial territories in Trentino, those that can best support the creation of territorial systems of scenario that include various VCs. Undoubtedly, the VC administrations were the main conversation partners for the designing of territorial scenarios. These scenarios could become more effective if they were inserted into an inclusive approach of governance, open to including the various territories/actors of the VCs in order to share actions and strategies leading them in the direction of whatever common scenario they wish to design and implement.

In this process, scenarios are anchored in a set of shared criteria[1] that can be traced back to inclusive processes of local action that the VCs had to perform according to the formulations of the Community Territorial Plan (CTP). These processes are founded, in the first place, on active local territorial policies defined by each VC's territorial governance.

The scenarios that the research proposed intend to become meaningful during the process of making practical decisions. Hence the scenarios are constructed as actions are being performed in the give and take of institutional proceedings where evaluation methods are used in order to insert mid-term decisions into long-term reflections. In this way, territorial scenarios become meaningful systems of governance oriented to reducing the fragmentary nature of territorial planning through promoting multi-level and multi-sector policies (Albrechts, 2010; Alexander, 2009; Balducci, 2010; Healey, 2009; Hillier, 2010). The goal is that of designing and implementing proactive conservation policies of the environment/landscape/land-use, the key components of the proposed model for territorial-economic enhancement and development.

These territorial scenarios can lead to proposals for action addressed to the VCs, proposals that need to be constructed entirely and that are doubtlessly more complex to practise than the conventional development planning processes. These proposals for strategic actions for the design of territorial scenarios are very different from the models generally used to design the retail territorial planning which have dominated the expansive dynamics of the sector nationally ever since the first national reform. The integrated evaluation of land-use, landscape and retail enabled planners to define the principles for the designing

1 The criteria for VC governance correspond to the indicators of the TIE matrix, applied for checking the ongoing territorial dynamics and for identifying the territorial retail scenarios. See chapters 1 and 2.

of scenarios. They were aided in defining these principles by the emergence of several pathways to follow and themes to work on as they drafted the CTP for the retail territorial development (Figure 6.1):

- The relationship with the infrastructure system – networks of primary accessibility along with the requirements that there be *pedestrian and bicycle accessibility* and that retail activities be integrated with the public transportation network in order to reduce any growth in traffic or pollution from emissions;
- The relationship with the urban context – historical city/rural areas/high-quality natural areas – implies a *return to the cities* and a search for new supply models and formats characterized by a readiness for reuse and recycling rather than by of the consumption of territorial resources;
- The relationship with the territorial and local-landscape system leads planners to define a unified project for a *diversified mix of retail and other functions – diversification/integration/short supply chain.*

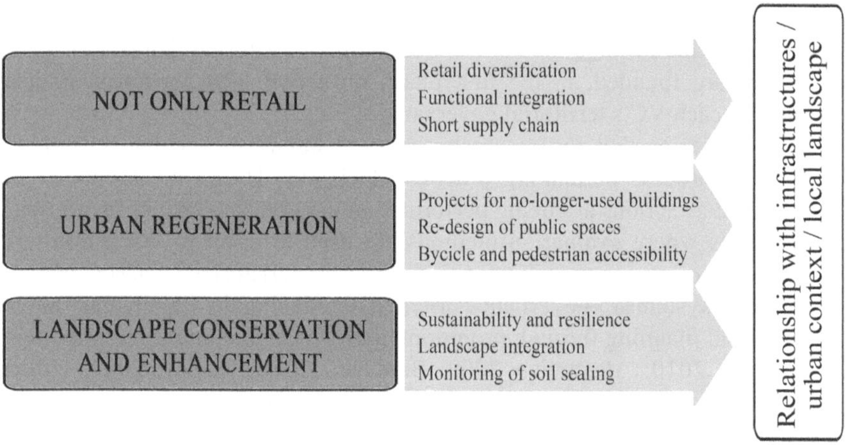

Figure 6.1 Principles for designing territorial retail development
Source: Author's elaboration.

Following the logic of these principles, the proposal calls for designing integrated actions of territorial enhancement and development of the three types of territorial scenarios (see Chapter 3) identified in five 'territorial systems' in the Province. These five macro-areas are characterized by their homogeneous cultural features, potential synergetic vocations for development, and intense flows of physical and functional relationships (maps 6.1 and 6.2).[2]

2 In our research, physical relations are considered to be the main connections existing among the VCs in terms of geomorphological, hydrographic and infrastructure

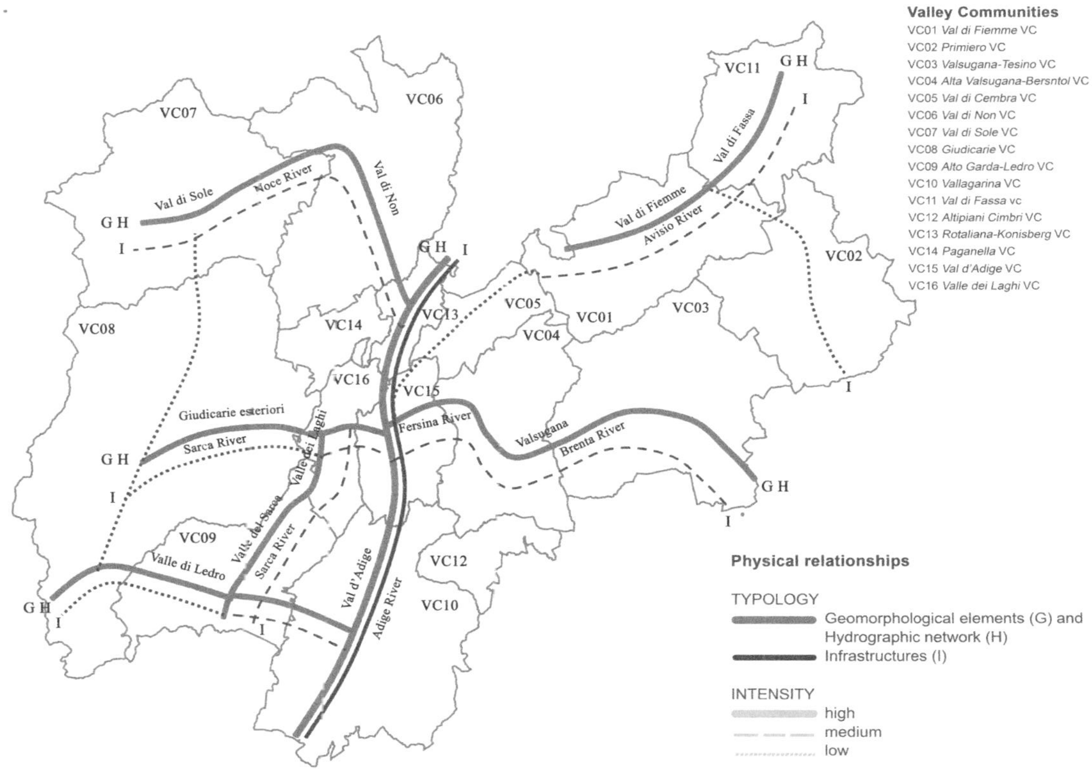

Map 6.1 **Physical relationships: The state of the art**

Source: Author's elaboration.

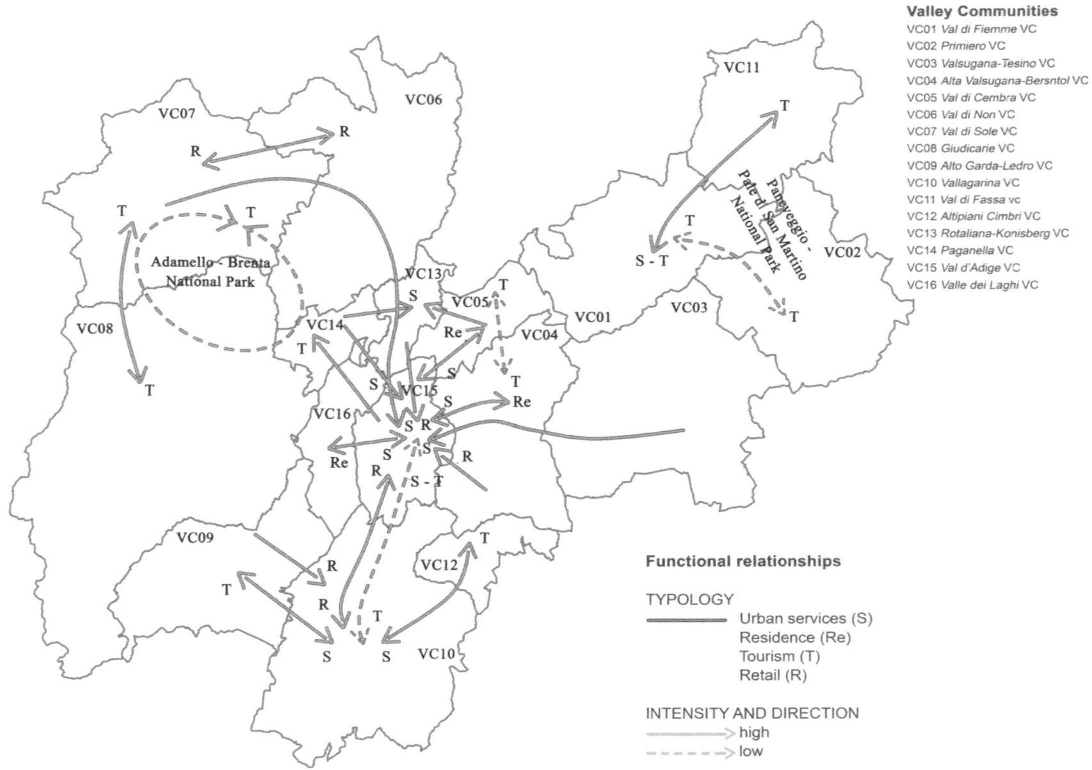

Map 6.2 Functional relationships: The state of the art

Source: Author's elaboration.

The following are the five 'territorial systems' that are characterized by their related systems of scenario. These five areas do not cover the entire Province. They do not involve all of the VCs, but they only include those areas – and the related scenarios – where planners foresaw concrete systematic interactions (see Map 6.3).[3]

- *The two-pole system along the Adige Basin* is composed of the Rotaliana, Val d'Adige and Vallagarina VCs and the municipalities of Trento and Rovereto. These 'territories' form a characteristic system because their development scenarios and the strategies related to them are clearly similar and because there are relationships that create a continuity that runs across the areas mainly in terms of physical characteristics (geomorphology, hydrography and infrastructures) and functions (with Trento as the primary pole of attraction for urban and especially administrative services for flows of people from the Vallagarina and Rotaliana VCs);
- *The western system* is composed of the Val di Sole, Val di Non and Giudicarie VCs. These all share a type of territorial scenario that is particularly outstanding for its exceptionally high-quality landscape and very attractive tourist systems. There are constant relationships among the three VCs, so much so that they form a real system, both physical (especially of the Val di Sole and Val di Non VCs) and functional (with Adamello Brenta Nature Park and a tourist system founded on ski facilities shared by the three VCs);
- *The eastern system* is composed of the Val di Fiemme, Val di Fassa and Primiero VCs. Like the western system, this system is unified by a development scenario where the local tourist system and landscape values play important roles (however, there are more differences among the territories than in the more homogeneous western system). These three VCs are also united physically and functionally, especially by their shared system of tourist-skiing offerings and shared Paneveggio and Pale di San Martino Nature Park, which is an important unifying element even though it is presently still a weak point;
- *The Valsugana system* is composed of the Alta Valsugana-Bersntol and Valsugana-Tesino VCs and it is marked by a strong and evident continuity

continuity. Functional relations are the main flows of people among the VCs in relation to four main attractions: services, residences, tourist resources and retail activities.

3 In this sense, according to our research, there is a first level of integration that occurs inside each inter-city administrative system represented by the VCs when the 'dominant' and 'prospective' scenarios are defined. There is a second level of territorial interactions that are related to flows of relationships among VCs, a level that configures potential territorial assets with 'variable geometry' represented by *systems of scenario*.

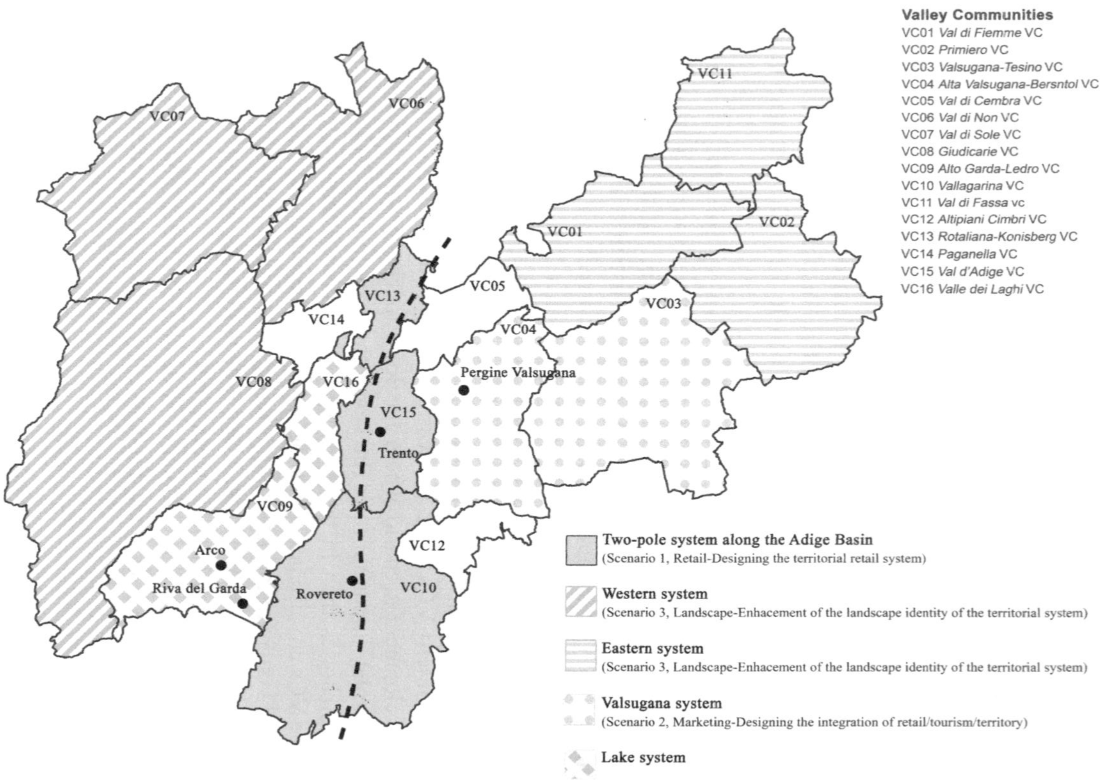

Map 6.3 Systems of territorial scenario
Source: Author's elaboration.

both physically (geomorphology and hydrography) and functionally (gravitation towards Trento for urban and retail services);

- *The lake system* is composed of the Alto Garda-Ledro and Valle dei Laghi VCs. This system is founded on the strong physical relationships, related to the morphology and landscape.

These five 'territorial systems' need to become spatial frames for effective forms of territorial governance. For this to happen, they need strongly design-oriented actions of governance that can help strengthen the potential relationships among the VCs by setting up *system strategies*. Thus the five systems of scenarios need to work towards initiating inter-sector actions for the enhancement of endogenous resources. In particular, the systems of scenarios would begin innovative actions for territorial development.[4] There are many potential actions that may be important for reinforcing the physical and functional relationships in some of the five 'macro-areas' (see Map 6.4). One of these, an outstanding example, is the strengthening of the bicycle-ready infrastructure connections that have been designed but not yet realized. This project for the extension of the bicycle network and the activation of a network of integrated mobility (bicycle-railroad, bicycle-vehicle) has become one of the strategic objectives for promoting a new tourist model for the Trentino.[5]

All in all, the general aim of the spatial policies for setting up scenarios has to do mainly with two interconnected lines of action:

- the enhancement of landscape, the environment and the land-use through the initiation of positive interactions among the resources in the various areas;
- the improvement of territorial networks and links, making the best of the opportunities that come from them, especially in relation to the future European strategy for an Alpine Macro-Region.[6]

4 The theoretic perspective assumed here recognizes that mechanisms of self-organisation must play leading roles as agents for meaningful and effective innovation in the construction of territorial development policies. Such a organisational perspective can be considered one among several extensive and heterogeneous experiments regarding different forms of associations beginning to re-emerge all over the world. Thus a new model of governance has emerged over the past ten years inspired by the principle of horizontal subsidiarity (Brunetta, 2015; Brunetta and Moroni 2012).

5 In this perspective are the goals of the Provincial Development Programme approved by the Provincial Council in December 2014.

6 Following a request from the European Council, the European Commission and the participating countries and regions are now proceeding to draw up an EU Strategy for the Alpine Region (EUSALP). See the first Consultation Document outcome of the Stakeholder Conference of the EUSALP (2014, December, Milan).

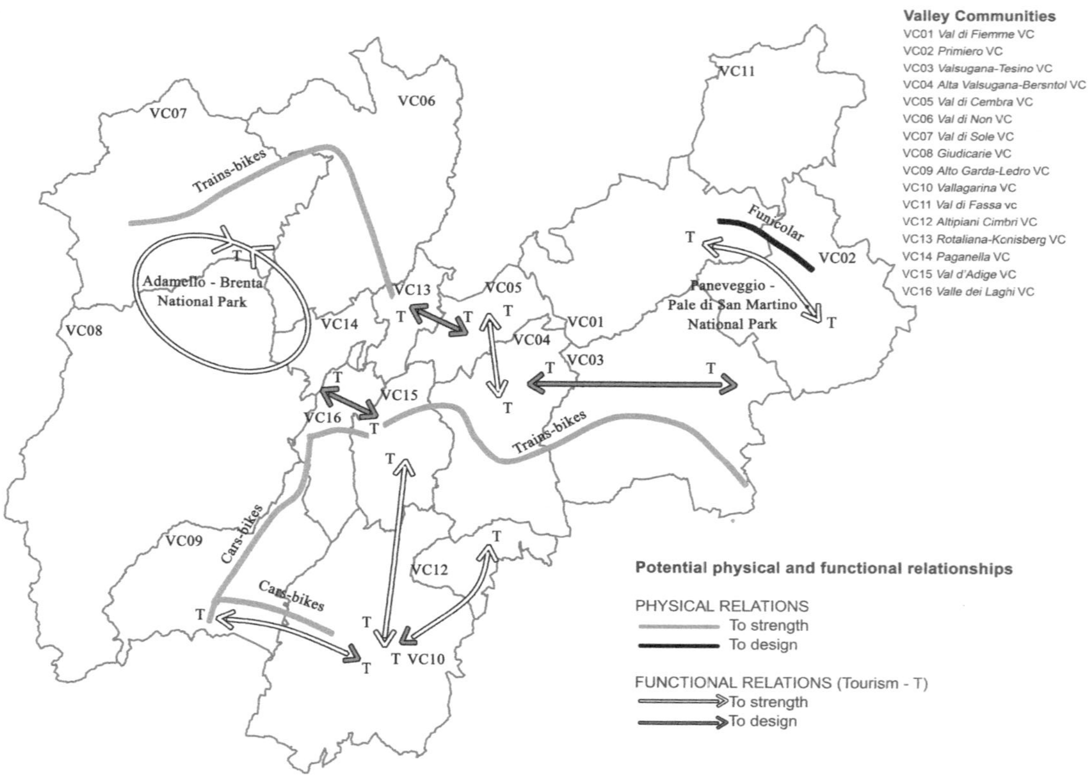

Map 6.4 Potential physical and functional relationships: The project

Source: Author's elaboration.

In consideration of these two lines of action, there is one especially meaningful goal – to strengthen efforts to take thorough care of the landscape,[7] understood in the broad sense. This can be done through integrated and inter-sector actions that enhance the excellent heritage of landscape resources and that renew urbanized areas. This can help meet the multiple objectives of economic development that reaches out towards the sustainability of the development processes. This strategy of territorial and environmental enhancement outlined in the systems of scenario is founded on a model of development that depends on the green economy,[8] and is oriented to reducing the consumption of natural resources through helping to increase energy efficiency, eliminate local pollution and improve the quality of life. Thus the specific characteristics of the territories can become components inside a new paradigm of development. They can promote the quality of space and the development of the economy by starting out from the potential territorial environmental and economic synergies that would be designed and implemented in the five 'territorial systems' in Trentino. This would not so much stop the transformation processes as much as orient them towards parameters of sustainability and urban resilience.[9]

6.2 Place-based Regional Policies in Trentino

The systems of scenario identified need to be put into operation. Therefore our research has proposed a form of management and a set of rules to promote the multi-level processes of governance in the five emerging macro-areas (our second and third spheres of innovation). We have detected potential processes of complementary and cooperative actions between several areas of the VCs and so we have proposed the designing of what can be called Territorial Units of Cooperation (TUC). These do not necessarily coincide with the administrative boundaries of the VCs, but they come out of the agreements among areas that share the same aims of renewal and development with one another.

The TUCs are therefore conceived of as *systems of territorial governance* with variable geometry. In other words, they are founded on the autonomous

7 The growing relevance of landscape has been taken for granted ever since the European Landscape Convention (ELC, 2000, ratified in Italy in 2006), which recognized the shared significance of landscape as a common good and basis for the identity and welfare of populations. Despite this, landscape topics have a hard time being considered entirely as components in all respects that cannot be ignored in the implementation of policies for territorial-economic enhancement and development.

8 In this perspective, see: Barbier (2010) and UNEP (2009).

9 The concept of resilience has been recently discussed in the urban and regional planning sector, becoming part of the vocabulary of spatial planning and entering debates in planning theory and practice. See: Davoudi et al. (2012); Pickett et al. (2004); Walker and Salt (2006); White and O'Hare (2014).

establishment of networks of structures/actors/resources in a given area. Inspired by the international experiences of the Business Improvement Districts,[10] this is a type of cooperative urban management that is capable of promoting diversified actions leading to local development and enhancement in Trentino's territories. The TUCs do not take the place of the VCs, but are ready to go into action when local actors begin to start them up. Their goal is to make any particular strategic action effective for local development by involving a variety of institutional actors in the process (for example, municipalities included in various VCs).

Therefore TUCs should be understood as 'place-based forms' of institutional cooperation, forms marked by a strong territorial-oriented[11] relational component for the triggering of potential actions for territorial enhancement. Thus there has been an attempt to promote a kind of 'collaboration through pacts' among the VCs. These agreements can be implemented for designing specific territorial actions for enhancement and development. Here the forms of interaction are hardly structured at all from the formal (institutional) point of view, but they are bound to specific territorial policy objectives that are meant to be achieved. Territorial Pacts, one of the forms introduced by European Union Structural funding to promote the cohesion regional policy, are examples of these type of interactions.

This institutional perspective has turned out to be in harmony with the recent national reform law of the territorial public bodies,[12] which highlights the current relevance and the potentialities of similar forms of territorial governance, especially the clustering of cities in order to implement joint policies of enhancement and development. The future institutional arrangement, which was initiated at the beginning of the year, calls for 'consolidations' and 'unions' of municipalities for outlying areas. In this perspective, the intermediate level of APT territorial government could be strengthened in the VCs, where conditions

---

10 BIDs are self-organized and self-funded forms of urban management, authorised by government, promoted and managed by private operators to determine the economic, social and territorial growth of an area through the promotion of retail. As Hoyt (2004: 367) states, 'although there is no standard definition, most scholars and practitioners describe business improvement districts as self-imposed financing mechanism implemented by business and property owners for local improvements, specially the enhancement of public services'. For an outline of the origin and evolution of Business Improvement Districts (BIDs) in the United States and Town Centre Management (TCM) in Great Britain, see: Dixon (2005); Morcol et al. (2008); Segal, (1998).

11 In a perspective in which economic behaviour and interpersonal activities are explained as a process that is influenced by the characteristics of places, UTCs can be interpreted as a comprehensive approach to community development in terms of urban regeneration, social inclusion and economic growth. See Brunetta and Caldarice (2014).

12 Law n.56. of 7 April 2014, Resolution for metropolitan cities, for provinces, and the union and merger of municipalities. The aim of the reform is to adapt the territorial administrations of the Country to the principles of 'subsidiarity', and 'adequacy' in the territorial services supply.

allow, starting out with encounters about strategic inter-city actions to promote for the designing of prospective scenarios.

To support the designing of systems of scenarios, the APT adapted the criteria of the TIE methodology, transforming them into new rules for the governance of the VC retail territories. The Provincial Council passed Resolution 1339 of 1 July 2013, which transformed the principles and criteria of TIE into regulatory guidelines for orienting territorial development planning, with a particular focus on the retail development in the built-up areas in the VCs.

The conditions and parameters identified by this research for the designing of scenarios have become a multi-faceted set of regulations concerning the self-evaluation in the CTP.[13]

When Resolution 1339 adopted the TIE parameters, it introduced in the Province a new rational paradigm in territorial planning, one that was no longer founded on the quantitative parameters of retail-development planning. Instead, it was addressed to promoting the quality of territorial development processes. In respect to the potentialities and the special features of each VC, the resolution was meant to link the quality of the territorial retail development with environmental sustainability. Such sustainability can be articulated in terms of enhancing territorial resources, containing environmental pollution, and integrating other sectors with the landscape-related context of accessibility. In particular, the principles of this approach for the governance and planning of territorial retail scenarios are the following:

- the promotion of the role of historical city centres as attractions;
- the definition of criteria for the 'recycling' and renewal of abandoned pieces of the territorial heritage (no-longer-used areas and buildings);
- the limitation of the dynamic of soil sealing.

The strong point for putting the reform into action is the autonomy of the VCs in making decisions. The VCs must apply the TIE set of qualitative criteria to discuss and design the territorial retail development scenarios in the CTP currently under development.

6.3 Open Questions and Working Perspectives

In this book so far, we have discussed our response to the needs for new knowledge and technical skills expressed by the APT when we began our research programme, needs to determine guidelines for innovating provincial planning. The results that we have reached so far and are still being experimented have

13 In the Provincial Resolution n. 1339/2013, the TIE self-evaluation must be applied in order to promote the improvement of the development processes in the territorial planning.

raised several general questions about the ongoing processes of institutional renewal being tried out in Trentino territories today. These process, as we have seen, have introduced new orientations for institutional action now being taken, action that is aimed at reinforcing, on the one hand, the integration of the territories (Province/VCs/municipalities) and, on the other hand, the integration of policies (land use, environment, landscape-related and economic). Some open questions and prospective for further work arise from these two themes of work: the integration of territories and of policies.

The first issue regards the prospective for governance that had been introduced, which brought on beneficial processes in several VCs, where resources/actors were networked in the effort to design the opportunities for enhancement and development defined by the territorial scenarios that emerged from the TIE. On the whole, these dynamics demonstrate that a new logic of action has emerged in several administrative areas in Trentino, a logic that can easily be combined with the TUCs proposed by the research. The overall aim is to re-compose new arrangements of spatial organization, where territorial retail development could become the spur for new processes of regeneration if linked up with other leading functions of the territorial economy, that is agriculture and tourism.

Arising out of autonomous unions of municipalities that do not necessarily coincide with the administrative borders of the VCs, the governance of the TUCs could foster local administrative patterns of behaviour that could develop and project a common territorial scenario, one in which material and non-material resources are shared and the strategies and investments that are needed are put into a network.

Enacted on the local scale with the approval of the Provincial Resolution, the TIE process, in several cases, is accompanying the work of territorial institutional actors (VCs and municipalities) with autonomous initiatives of cooperation. This phenomenon leads us to question whether the VCs are institutional spaces that are effective in acting competitively from the territorial point of view. In fact, the dimension of the territorial systems identified with the systems of scenario could make more 'room' for new territorial aggregations, defining coherent and effective strategies independent of the administrative perimeters of the VCs. Here the passage is clear – and surely complex to construct – to a form of cooperation among territories structured through agreements that are reached in order to manage specific territorial government aims (infrastructures, services, environment and landscape).

The second issue regards the leading role that urban areas play as actors fundamental for the effective activation of the strategies defined on the regional level. Current trends and crises in the economy and finance that we are going through call for new paradigms for the analysis and planning of urban phenomena. They call for alternative models of territorial development inspired by the principles of 'resilience', 'sustainability', 'responsibility' and 'subsidiarity'. In the specific case of activating the territorial retail development scenarios put forward by TIE, cities are delegated with implementing renewal and enhancement policies for historical city centres. The historical city centre is the place where tourism and retail can combine most efficaciously in order to create effective development scenarios both in terms of economic redevelopment and in terms of safeguarding landscape values, thus avoiding any additional processes of land-use consumption associated with new settlements in suburban areas.[14]

In particular, the city is again becoming a great attractor of new sophisticated settlement proposals that abandon the traditional retail formats that have been indifferent to their territorial contexts. Instead, these proposals envisage the reuse of city areas and buildings. In this case, retail can become the engine of urban regeneration, an attraction for initiatives for enhancement and requalification. As opposed to what happened in the preceding decade, the new retail activities are more and more oriented towards seeking out the highest level of spatial, formal, typological and functional integration with their urban settlement contexts. Furthermore, the economic crisis contributes to a readjustment in activities and in consumption that has shrunken the field of action of retail initiatives. Furthermore, the crisis contributes to the emergence of cultural models of consumption that are inspired by principles of environmental sustainability towards urban resilience, which require the planning of new retail territorial development (Barata-Salgueiro and Erkip, 2014; Fernades and Chamusca, 2014; Vale, 2013).

The third issue focuses on the prospective for planning that has been introduced. This prospective radically changes the established approach to urban retail planning and policies. This new approach highlights the territorial opportunities that could emerge when forms of integration among the policies of different sectors are implemented, making the competitive characteristics of the areas emerge. The regulations for territorial government, especially in Italy, are changing. It is fundamental to integrate the various sector-based policies. Integrating these policies is the only way to attempt to start to enhance, develop, maintain and safeguard premium areas. We are going through a completely new phase of territorial development and transformation without 'growth', a phase that requires that actions be made for the 'recycling' and requalification of what already has been urbanized. Such actions can be sustained by a radical cultural change in the approaches to designing territorial policies, improving the process of integration of territories/actors/resources, which can be the sources from the new 'topics' for the work needed on the territorial planning and urban project.

Here, both technical expertise and the inclusive capability of the public actor come back to play a decisive role in the building up of a political process (Bianchetti and Balducci, 2013; Friedmann, 2011). Technically, the TIE criteria have become the framework for public discussion on enhancement strategy. Politically, public actors are not immune to the rise of conflicts and risks for the actions, but they are doubtlessly full of promise in terms of institutional

14 See the Provincial Development Programme 2014– 2018.

learning because of the quality and transparency of the processes that have been implemented.

References

Albrechts, L., 2010. How to Enhance Creativity, Diversity and Sustainability in Spatial Planning: Strategic Planning Revisited. In: M. Cerreta, G. Concilio and V. Momo (eds), *Making Strategies in Spatial Planning*. Dordrecht: Springer, pp. 3–25.

Alexander, E.R., 2009. Dilemmas in evaluating planning, or back to basics: What is planning for? *Planning Theory & Practice*, 10(2): 233–44.

Balducci, A., 2010. Strategic Planning as a Field of Practices. In: M. Cerreta, G. Concilio and V. Momo (eds), *Making Strategies in Spatial Planning*. Dordrecht: Springer, pp. 47–65.

Barata-Salgueiro, T. and Erkip, F., 2014. Retail planning and urban resilience: An introduction to the special issue. *Cities*, 36: 107–11.

Barbier, E.B., 2010. *A Global Green New Deal. Rethinking the Economic Recovery*. Cambridge: Cambridge University Press.

Bianchetti, C. and Balducci, A. (eds), 2013. *Competenza e rappresentanza*. Rome: Donzelli.

Brunetta, G., 2015. The Contractual Communities' Contribution to Cultural and Natural Resource Management. In: R. Gambino and A. Peano (eds), *Nature Policies and Landscape Policies: Towards an Alliance*. Dordrecht: Springer, pp. 233–42.

Brunetta, G. and Caldarice, O., 2014. Self-organisation and retail-led regeneration: A new territorial governance within the Italian context. *Local Economy*, 29(4–5): 334–44.

Brunetta, G., and Moroni, S., 2012. *Contractual Communities in the Self-organizing City: Freedom, Creativity, Subsidiarity*. Dordrecht: Springer.

Davoudi, S., Shaw, K., Haider, L.J., et al., 2012. Resilience: A bridging concept or a dead end? *Planning Theory & Practice*, 13(2): 299–333.

Dixon, T., 2005. The role of UK retailing in urban regeneration. *Local Economy*, 20(2): 168–82.

Fernades, J.R. and Chamusca, P., 2014. Urban policies, planning and retail resilience. *Cities*, 36: 170–77.

Friedmann, J., 2011. Epilogue: Citizen Planners in an Era of Limits. In: J. Friedman, *Insurgencies: Essays in Planning Theory*. New York: Routledge, pp. 231–38.

Healey, P., 2009. In search of the 'strategic' in spatial strategy making. *Planning Theory & Practice*, 10(4): 439–57.

Hillier, J., 2010. Post-Structural Complexity: Strategic Navigation in an Ocean of Theory and Practice. In: M. Cerreta, G. Concilio and V. Momo (eds), *Making Strategies in Spatial Planning*. Dordrecht: Springer, pp. 87–97.

Hoyt, L., 2004. Collecting private funds for safer public spaces: An empirical examination of the business improvement district concept. *Environment and Planning B: Planning and Design*, 31: 367–80.

Morcol, G., Hoyt. L. and Meek, J.W. (eds), 2008. *Business Improvement District: Research, Theories and Controversies*. New York: CRC Press.

Pickett, S.T.A, Cadenasso, M.L., and Grove J.M., 2004. Resilient cities: meaning, models, and metaphor for integrating the ecological, socio-economic, and planning realms. *Landscape and Urban Planning*, 69: 369–84.

Segal, M., 1998. A new generation of downtown management organization. *Urban Land*, 4: 86–9.

United Nations Environment Programme, 2009. *Cities and Green Building. In the Transition to a Green Economy*. UNEP brief.

Vale, L.J., 2013. The politics of resilient cities: whose resilience and whose city? *Building Research & Information*, 42(2): 191–201.

Walker, B. and Salt, D., 2006. *Resilience Thinking: Sustaining Ecosystem and People in a Changing World*. Washington, DC: Island Press.

White, I. and O'Hare, P., 2014. From rhetoric to reality: which resilience, why resilience, and whose resilience in spatial planning? Environment and Planning C: Government and Policy, 32: 934–50.

Index

accessibility 9, 26, 33, 65, 67, 69–70, 72, 86–7, 91–2, 97–101, 106, 110, 119
  level of 65, 70, 86–7, 91–2, 101
  pedestrian and bicycle 110
  provincial 33, 65, 69, 70, 72, 100
  territorial 9, 26, 33, 67, 97–9, 101, 119
actors
  institutional 2–3, 75, 107, 118
  local 6, 8, 24, 118, 121
  public 18, 22, 121–2
  VCs and municipalities 8, 26, 109, 120
agreements 8, 14–16, 18, 67, 81, 117–18, 120
  cooperative 8, 14, 117–18, 120
Alexander, E.R. 1–2, 10, 13–15, 18, 28–9, 109, 122
alternatives
  cluster of 58, 77–8, 87, 89
  decision 3, 77–8
  evaluation of 75, 78–81, 84
Analytic Network Process 9, 26, 75–8, 80–86, 91–4, 96
  ANP, *see* Analytic Network Process
APT, *see* Autonomous Province of Trento
Archibugi, F. 14–15, 28
attractiveness 9, 27, 58, 65–7, 69–70, 72, 74, 97–105, 108
  index of 9, 27, 65–7, 69–70, 72, 98–9, 101–5, 108
  values of 58, 65, 74, 97–101, 104–5
Autonomous Province of Trento 2–4, 7, 13, 22, 26, 32, 54, 118–19

citizen 16–17, 78, 122
  citizenship 2
communicative planning 1–2, 16, 26, 29
Community Territorial Plans 6, 8, 24
criteria
  evaluation, *see* evaluation
  for territorial scenarios 7, 23–4, 31, 86
  meta-criteria 16, 26
  multi-criteria analysis 9, 19, 24, 75, 85, 94–6
  of decision 3, 14–15, 18–19
  shared 5, 7, 14, 18, 25, 109
  TIE, *see* TIE
CTPs, *see* Community Territorial Plans
cumulative territorial effects 17, 19, 20

decision-making
  model 75, 77–8, 95
  process 1–2, 4, 8–10, 13–15, 17–19, 21–5, 28, 78, 86, 93, 107

evaluation 1–7, 9–11, 13–29, 31–3, 36–41, 44–5, 49, 53, 57, 64, 75–6, 81, 85–8, 90, 95–8, 102, 105, 107, 109, 119
  alternatives, *see* alternatives
  approach 3, 14–16, 18, 26, 32, 75
  as teaching and learning 2, 15
  Community Impact Evaluation 11, 16, 29
  criteria 9, 18–19, 27, 76, 87, 91–2
  design of 3, 10, 16
  in spatial planning 2–3, 5, 9–11, 13–14, 16, 18, 21–2, 24, 28–9, 75, 109
  matrix 9, 27, 38–9, 40–41
  meta-evaluation 14, 32
  methodology 4, 13, 15, 16, 18–20, 26, 109
  process of 3–4, 6, 13–14, 16–19, 22, 25, 26, 88–90
  rationality 3, 14

scale 26–7, 38, 44, 64, 86
self-evaluation 7, 119
smart 2, 3, 9
strategic 11, 13–16, 85, 87
Territorial Integrated, *see* TIE
evaluation themes
land use 20, 26, 33–4, 44, 46–7, 59–60, 109,
landscape 7, 19–20, 26–8, 32–3, 35, 37, 40–42, 46, 48–50, 62, 86, 91–2
local governance, *see* governance
territorial retail supply, *see* retail
tourism 5, 6, 20, 24, 33–4, 37–8, 44, 46–7, 49, 57, 101, 108

Forester, J. 2, 10
Friedmann, J. 1, 10, 121, 122,

GLAs, *see* gross leasable areas
Guba, E.G. and Lincoln, Y.S. 1, 10, 15, 25, 28
governance
level/multi-level of 2–3, 65, 117
local 7, 20, 27, 33, 35, 46, 48, 63–5, 93, 109, 119
processes of 9, 13–14, 41, 107, 115, 117
systems of 6, 109, 115, 117
territorial 7, 9, 11, 14, 21–2, 26–7, 41, 45, 74, 86, 107, 109, 115, 117–18, 120, 122
gross leasable areas 4, 7, 9, 17, 33, 53, 86, 97
guidelines 7, 24–7, 39–40, 42, 86, 119
planning 25–7, 86, 119
regulatory 7, 24, 119
strategic 39–40, 42

Healey, P. 2, 10
Hillier, J. 3, 10, 17, 28, 109

indicators 9–10, 22, 27, 34–5, 38–42, 44–7, 49, 63, 75, 77, 83, 87, 90–93, 97–8, 101–2, 109, 119
matrix of 9, 27, 34–5, 38–41, 109

institutional
actors, *see* actors
capacity-building 5, 8, 14
design 2–3, 5–6, 10, 14, 19, 28
framework 14, 25–6
innovation 1, 5, 8–9, 32, 107
knowledge, *see* knowledge
learning, *see* learning
renewal 2, 120
integration
design 27, 67, 108
landscape 27, 36–7
policies 14, 107, 120–21
functional 9, 24, 65, 67, 69–70, 72, 87, 101, 121
retail 5, 9, 27, 31–2, 34, 37, 40, 43, 47, 65–6, 69–70, 73, 107, 110
territories/actors/resources 2–3, 16, 36, 40, 67, 69–70, 86, 93, 108, 120–21

Khakee, A. 1, 10–11, 14, 15, 18, 29
knowledge
shared 16
technical 14, 16, 26, 119
institutional 3, 26
paradigms 4, 7, 32
TIE, *see* TIE
usable 14

learning
collective 2
institutional 2, 7, 15–16, 19, 22, 26, 41, 122
proactive institutional 16, 7
process 1, 14–16, 19, 22
Lichfield, N. 1, 11, 13, 14, 16, 18, 29
Lidblom, C.E. 14, 29
local economy 20, 26, 32, 56, 74, 122

March, J.G. and Olsen, J.P. 15–16, 29
marketing 27–8, 32, 66–7, 69–71, 86, 91–3, 108
territorial 32
retail/tourism/territory 28, 67, 69–70, 86, 91

Miller, D. and Patassini, D. 1, 11, 13, 18, 28–9, 76, 95
multivariate statistical
analysis 98, 101, 106
method 97
technique 98

Provincial Resolution 1339 7, 24, 44, 119
Provincial Territorial Plan 4–5, 13, 25–6, 37–42, 61, 87
PTP, *see* Provincial Territorial Plan

rationality 1–3, 14–15, 18–19, 21, 23, 26, 109
approach 1–2, 15
in evaluation, *see* evaluation
limited 2, 15
paradigms 2, 21
TIE, *see* TIE
weak 3
regional development 2, 5–6, 8, 14, 32
paradigms 5
policies 8, 14, 32
integrated design in 2, 5
regional planning 4, 6, 8, 13–14, 19, 29, 31, 85, 96, 117
approach to 6, 13
process of 4, 14
retail, *see* retail
resilience 117, 121–3
retail 2, 4–9, 13, 17, 19–20, 22–8, 31–41, 43–51, 53–7, 64–74, 76, 85, 91, 93, 101–2, 104–10, 113, 115, 118–22
development 4, 7–8, 13, 17, 19–20, 24, 26, 31–2, 67, 74, 76, 85, 107, 110, 119–21
integration, *see* integration
planning 4–5, 8, 19–20, 23, 26, 28, 32, 37, 50, 72, 93, 109, 119, 121–2
policies 5–6, 13, 17, 24, 26, 31, 37, 107
scenario, *see* scenario(s)
settlement 5, 7, 17, 19–20, 24, 44, 65, 69
supply 20, 24, 33–4, 36, 40, 44–7, 49, 53, 56, 65, 66, 69, 71–2, 93, 102
system 20, 27, 32, 37, 40, 53, 54–6, 64, 66–7, 69, 72, 102, 108

Saaty, T.L. 75–8, 80, 82, 85, 89, 91, 94–6
scale
macro 27, 34–5, 38, 40, 44–5, 47, 53, 59, 64, 86–8, 91–3
micro 27, 34–5, 38, 40–41, 44–6, 53, 56–8, 60, 64–5, 77, 86–8, 91–3
of evaluation, *see* evaluation
scenario
design(ing) 2, 5–8, 17, 24, 32, 37, 108–9, 119
dominant and prospective territorial 22, 25, 66–7, 69–70, 102, 105, 108, 113
emerging territorial 22, 25, 27, 41
local development 2, 6, 9, 13–14, 19, 22, 25, 27, 38, 65, 69–70, 72, 93, 107, 113, 121
long-term 92, 108
mid-term 108
system of 107–9, 113–15, 117, 119–20
territorial retail 2, 9, 17, 19, 24, 26–8, 33, 36–9, 41, 65, 68, 86, 93, 102, 105–9, 119, 121
soil sealing 20, 40, 42, 73, 119
stakeholders 13, 15–16, 19, 25–6, 75, 86, 88, 107
subsidiarity 3, 16, 115, 121, 115, 118
horizontal 115
sustainability 5, 7, 25–6, 37, 40, 42, 86, 94, 117, 119, 121
environmental 94, 119, 121
of development processes 117, 119
urban 26,
SWOT analysis 19, 24, 26, 49, 79, 85, 86, 87, 93, 105

territorial development
planning and design of 22, 24
polycentric 5, 107
Territorial Integrated Evaluation, *see* TIE

Territorial Units of Cooperation 117, 118, 120
  as systems of territorial governance 117
TIE 32–4, 36, 38–42, 45, 49, 53, 65, 67, 76–7, 85–6, 89, 93–4, 105, 107–9, 119–21
  application of 5, 7, 9, 16, 24, 28, 102
  approach to 6, 16, 19–20, 25
  criteria 6–9, 24, 27, 107–9, 119, 121
  matrix 9, 27, 40, 49, 53, 65, 67, 109
  methodology 2, 4, 7–9, 13–14, 19–22, 24, 32, 41, 77, 85, 93, 107, 119
  process 3, 6, 14, 17, 22, 25–6, 32, 45, 86, 94, 120
  rationality 2–3, 26, 109
  self-evaluation 7, 119
TUCs, *see* Territorial Units of Cooperation
valley communities 4, 7, 22–3, 26, 32, 37–8
value(s)
  attractiveness, *see* attractiveness
  cultural 64
  ecological 64
  identity-related 41
  landscape 4–6, 32, 36–8, 41, 64, 67, 73, 113, 121
  shared 28, 50
  social 15, 19
VCs, *see* valley communities